IET RADAR, SONAR, NAVIGATION AND AVIONICS SERIES 3

Series Editor: Professor E.D.R. Shearman

Weibull Radar Clutter

Other volumes in this series:

Volume 1	**Optimised radar processors** A. Farina (Editor)
Volume 3	**Weibull radar clutter** M. Sekine and Y. Mao
Volume 4	**Advanced radar techniques and systems** G. Galati (Editor)
Volume 7	**Ultra-wideband radar measurements: analysis and processing** L. Yu. Astanin and A.A. Kostylev
Volume 8	**Aviation weather surveillance systems: advanced radar and surface sensors for flight safety and air traffic management** P.R. Mahapatra
Volume 10	**Radar techniques using array antennas** W. Wirth
Volume 11	**Air and spaceborne radar systems: an introduction** P. Lacomme (Editor)
Volume 13	**Introduction to RF stealth** D. Lynch
Volume 14	**Applications of space-time adaptive processing** R. Klemm (Editor)
Volume 15	**Ground penetrating radar, 2nd edition** D. Daniels
Volume 16	**Target detection by marine radar** J. Briggs
Volume 17	**Strapdown inertial navigation technology, 2nd edition** D. Titterton and J. Weston
Volume 18	**Introduction to radar target recognition** P. Tait
Volume 19	**Radar imaging and holography** A. Pasmurov and S. Zinovjev
Volume 20	**Sea clutter: scattering, the K distribution and radar performance** K. Ward, R. Tough and S. Watts
Volume 21	**Principles of space-time adaptive processing, 3rd edition** R. Klemm
Volume 101	**Introduction to airborne radar, 2nd edition** G.W. Stimson
Volume 102	**Low-angle radar land clutter** B. Billingsley

Weibull Radar Clutter

Matsuo Sekine
and Yuhai Mao

The Institution of Engineering and Technology

Published by The Institution of Engineering and Technology, London, United Kingdom

First published 1990
Reprinted with new cover 2009

The Institution of Engineering and Technology
Michael Faraday House
Six Hills Way, Stevenage
Herts, SG1 2AY, United Kingdom

www.theiet.org

British Library Cataloguing in Publication Data
Sekine, Matsuo, 1945–
Weibull radar clutter
1. Radar
I. Title II. Mao, Y.H.
621.3848

ISBN (10 digit) 0 86341 191 6
ISBN (13 digit) 978-0-86341-191-5

First printed in the UK by BPCC Wheatons Ltd, Exeter
Reprinted in the UK by Lightning Source UK Ltd, Milton Keynes

Contents

Preface

Radar clutter is defined as the unwanted reflective waves from irrelevant targets. The amplitude statistics of clutter have been modelled by Rayleigh, log-normal, contaminated-normal, Weibull, log-Weibull and K-distributions.

During the past twenty years there has been a considerable growth of interest in various Weibull-distributed ground, sea, sea ice and weather clutter returns concerning false alarms and effective detection processes.

In the opening chapter, the reader is introduced to the derivation of Weibull distribution in a general way. Chapter 2 deals with radar clutter as statistics with Weibull distribution.

As with signal detection methods in clutter, parametric constant false alarm rate (CFAR) detectors in Weibull clutter and non-parametric CFAR detectors are dealt with in Chapter 3.

The suppression of radar clutter is a very important problem in modern radar technology. There are many papers and books discussing this problem. We do not discuss this problem in general, but discuss some special problems concerning Weibull clutter, these subjects being covered in Chapter 4 and the Appendixes.

This comprehensive work should prove invaluable, not only to radar engineers, but also to all who are in any way concerned with Weibull distribution.

During the preparation of the manuscript, we recorded, with deep regret, the death in November 1988 of Dr John Clarke of the Institution of Electrical Engineers and the Royal Signals and Radar Establishment of the United Kingdom, who encouraged us to write this book. We should like to dedicate this book to his memory.

The manuscript as prepared while Yuhai Mao was with the Department of Applied Electronics at the Tokyo Institute of Technology in Japan, and Matsuo Sekine was with the Department of Electromagnetic Theory at the Royal Institute of Technology in Sweden.

We had the benefit of lengthy discussions with a number of colleagues and friends in Japan, The People's Republic of China, and Sweden. We thank, in particular, Prof. Toshimitsu Musha of the Tokyo Institute of Technology, Prof. Yoshiwo Okamoto of Chiba Institute of Technology, Prof. Zhao-Da Zhu of Nanjing Institute of Aeronautics, Prof. Zai-Gen Fang of Beijing Institute of Technology, Prof. Staffan Ström of the Royal Institute of Technology, and the late Prof. Göran Lind and Prof. Gerhard Kristensson of Lund Institute of Technology, for their valuable suggestions.

Finally, the permission by the various journals to reproduce the material here is gratefully acknowledged.

Matsuo Sekine
Yuhai Mao

Introduction

There are two kinds of radar echo. One is the echo signal reflected from the targets which we are interested in, such as aircrafts, ships, guided missiles etc. The other is the echo signal reflected from unrelated bodies, such as land, clouds, rain, snow, sea waves, birds, insects and angels (atmospheric turbulence). However, 'useful' signal is meant simply in the relative sense. The useful signal for some applications may become the harmful signal in another case. For example, the echo signal from clouds is useful for weather radar, but is harmful for most other radars. In general, the background echo, which is termed 'clutter', is harmful for most radars. The useful target signal is often embedded in the clutter. Therefore, how to suppress clutter and to detect target signals from the clutter is one of the most important problems in modern radar technology.

First, we have to study the statistical properties of the clutter echo. These are very complex. They not only depend on the type of background, but also on the parameters of the radar, such as resolution, wavelength, polarisation etc. Since the environment will change with time, the clutter signal also exhibits non-stationary statistics. This adds more complexity to the study of statistical properties of clutter.

Thus we have to search for methods for clutter suppression. Since the statistical properties of clutter for different environments are different, the methods of suppression will be also different for different types of clutter. For example, an ordinary MTI system is sufficient for the suppression of ground clutter, but is ineffective for suppression of weather clutter, sea clutter, or other moving clutter. Adaptive MTI systems are effective for most forms of clutter and they can track the variation of the clutter spectrum. However, a good adaptive MTI system design also implies knowledge of the statistical properties of the clutter.

There are many kinds of characteristics which describe the properties of radar clutter. The principal ones are relative intensity, amplitude distribution, autocorrelation functions, in time and in space, spectrum, polarisation properties etc. All depend on the specification of the radar, especially the wavelength of the resoluton cell (i.e. the pulse width and the beamwidth) of the radar, and grazing angle, on one hand; and on the type of the clutter, such as land, weather, sea and chaff etc, on the other.

We often use the backscatter coefficient σ_0 to represent the relative intensity of the clutter. The radar cross-section of the clutter can be calculated from this coefficient and the resolution cell of the radar. It can be calculated for low grazing angles from

$$\sigma = R\Delta\theta\left(\frac{c\tau}{2}\right)\sigma_0 \tag{1}$$

where R is the distance from clutter to radar, $\Delta\theta$ is the azimuth beamwidth of the radar, c is the velocity of light, and τ is the pulse width of the radar. Decreasing the size of resolution can decrease the intensity of clutter signal.

The mean backscatter coefficient depends mainly on the type of clutter. The largest is for land clutter, next comes sea clutter and then weather clutter. In the case of land clutter, it depends on the type of terrain. This ranks as follows: cities, mountains, small house districts, wooded hills, open woods, cultivated land and desert. The median backscatter coefficient increases somewhat with frequency for most terrain types, but usually not more than linear with transmit frequency. The median backscatter coefficient increases about linearly with depression angle from 0·5° to 10° below the horizontal.

Since the clutter signal is not a steady signal, it fluctuates with time and space. Therefore, it is better to consider the clutter signal as being a random sequence, and to study its statistical properties.

The most important property may be the amplitude distribution of clutter. This has been described using Rayleigh model assumptions for many years. The reason for this is the mathematical simplicity of computation and the lack of knowledge of real clutter. However, as the size of the resolution cell of the radar decreases, the clutter distribution develops a larger tail than the Rayleigh distribution. Consequently, if the Rayleigh model is still used, the predicted false-alarm probability will be incorrect, and the CFAR detector based on this assumption cannot maintain the false-alarm rate constant in real clutter conditions.

Recent investigations of natural clutter characteristics have shown that the clutter echo can be approximated by a Weibull distribution. It is noteworthy that the Rayleigh distribution is a special case of the Weibull distribution family. The truth of the Weibull distribution reported first in land clutter returns as seen by high-resolution radars, and the skewness of the Weibull distribution was shown to increase as the radar depression angle was decreased [1]. Recently, Weibull distribution has also been observed in weather clutter [2] and sea clutter [3]. These facts are the principal motivation for the use of the Weibull clutter model.

Weibull distribution was proposed by a Swedish Professor, Waloddi Weibull, in 1939 [4] to establish a statistical theory of the strength of materials. Up to 1977, more than 1000 books and papers had been published on the theoretical properties of the distribution, its use for the statistical analysis of test data, and its practical applications [5].

The Weibull distribution is given by

$$p(x) = \begin{cases} \dfrac{\eta}{\sigma}\left(\dfrac{x}{\sigma}\right)^{\eta-1} e^{-(x/\sigma)^\eta}, & x > 0, \sigma > 0, \eta > 0 \\ 0, & \text{otherwise} \end{cases} \tag{2}$$

where σ and η are the scale and shape parameters, respectively. For $\eta = 2$ we get the Rayleigh distribution

$$p(x) = \frac{2x}{\sigma^2} e^{-x^2/\sigma^2} \tag{3}$$

Some examples of the Weibull distribution are plotted in Fig. 1, where $\sigma = 1$. We can see from this Figure that the Weibull distribution with $\eta < 2$ has a larger tail than the Rayleigh distribution ($\eta = 2$).

However, clutter as seen from a fixed point varies both in time at a given range and spatially. Spatial and temporal distributions are generally different. This means that the data collected from a fixed area has a distribution which differs from that collected from different regions. In practice, the former corresponds to the tracking radar, and the latter corresponds to the search radar with a scanning antenna. For land clutter, temporal distributions (observed spatial grid or footprint fixed) are usually considered to be Rician.

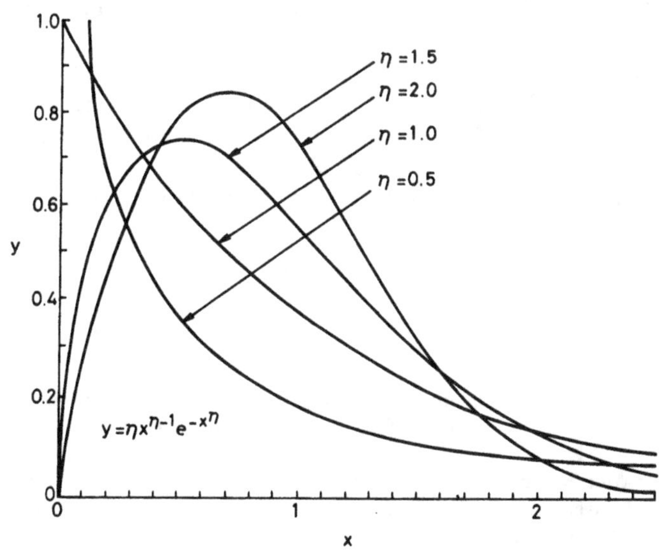

Fig. 1 *Weibull distribution function*

But some authors have reported that the temporal distributions for trees between 9·5 and 95 GHz has log-normal shape and sometimes Weibull shape. Reported spatial distributions for land extend from the Rayleigh to large standard deviation log-normal and Weibull distributions.

Since Weibull distribution is a very flexible distribution, one can change its shape by means of changes in the shape parameter. So it can be fitted to many different types of clutter. In fact, clutter reflected from a specified area may change its shape parameter not only with the resolution of the radar, but also with time for a given radar. For example, sea clutter will change its shape parameter as the wind speed changes [3]. In general, the shape parameter will decrease as wind-speed increases. So we can use the Weibull distribution with a wind-dependent shape parameter conveniently to describe sea clutter. This is the reason why the Weibull distribution has been widely accepted.

The clutter signals are correlated in space, both in range direction and in azimuth direction. In general, the correlation area corresponds to the resolution cell. This means that the correlation time in the range direction corresponds to the pulse width; and in the azimution direction it corresponds to the beamwidth. However, in the azimuth direction, since the antenna scanning time for one beamwidth will occupy 10 to 20 pulse repetition periods, the clutter may fluctuate in this time period. Therefore, the correlation time of clutter in the azimuth direction is often less than the time needed for the antenna scanning the beamwidth.

It is better to describe the fluctuation of the clutter in time with its spectrum. The spectrum of the clutter can not only describe the speed of fluctuation but also can reveal the Doppler frequency of the moving clutter. This is very useful for the design of adaptive clutter cancellers in the frequency domain.

In this book we will begin with the introduction of the fundamental principle of the Weibull distribution and its applications. Then, we provide a summary of the measured data of land clutter, weather clutter and sea clutter, which can be modelled with the Weibull distribution. After describing the detection problems in Weibull clutter, we discuss the methods of suppression of Weibull clutter.

References

1 BOOTHE, R. R.: 'The Weibull distribution applied to the ground clutter backscatter coefficient', US Army Missile Command, Technical Report, RE-TR-69-15, AD A691109, 1969

2 SEKINE, M., MUSHA, T., TOMITA, Y., HAGISAWA, T., IRABU, T. and KIUCHI, E.: 'On Weibull distributed weather clutter', *IEEE Trans.*, 1979, **AES-15**, pp. 824–830.

3 FAY, F. A., CLARKE, J. and PETERS, R. S.: 'Weibull distribution applied to sea clutter', Radar 77, 1977, pp. 101–104.

4 WEIBULL, W.: 'A statistical theory of strength of materials'. I.V.A.-Handl. No. 151, 1939.

5 WEIBULL, W.: 'References on the Weibull distribution'. FTI A report, A20:23, Aug. 1977.

Chapter 1

Fundamentals of Weibull distribution

Weibull distribution was suggested by Waloddi Weibull, a Swedish professor, in 1939, to explain the well known but unexplained facts that the relative strength of a specimen decreases with increasing dimensions and that its bending strength is larger than its tensile strength [1].

This theory was based on the assumption that the strength is a stochastic quantity, which has to be specified by a distribution function with one or more parameters.

It was also assumed that this function is a property of the material, and that the previously mentioned size effects are reflected by changes in the values of the parameters of the given function.

Let T be the ultimate tensile strength of a bar or a wire of length $L = 1$, and $F(t) = \text{prob}(T \leqslant t)$ be its cumulative distribution function. Then the probability of failure at a load equal to t will be $F(t)$ and the probability of non-failure is equal to $1 - F(t)$. If now the length of the bar is doubled, it is evident that the probability of non-failure $1 - F_2(t)$ is equal to the probability that neither of the halves of the bar fails; that is

$$1 - F_2(t) = [1 - F(t)]^2$$

and, in general, for any arbitrary length L,

$$1 - F_L(t) = [1 - F(t)]^L \tag{1.1}$$

or

$$F_L(t) = 1 - [1 - F(t)]^L \tag{1.2}$$

Let us assume that $F(t)$ is the normal distribution function. Then eqn. 1.2 proves that $F_L(t)$ is not a normal distribution function; i.e. if the strength of a bar is normally distributed for a certain length, the normality will be definitely excluded for any other length.

From eqn. 1.1 it follows that

$$\log[1 - F_L(t)] = L \times \log[1 - F(t)] \tag{1.3}$$

Thus it is required that $\log[1-F(t)]$ be a function of t; i.e.

$$\log[1-F(t)] = -g(t) \tag{1.4}$$

The most simple two-parameter function is given by

$$g(t) = (t-a)/b \tag{1.5}$$

and the most simple three-parameter function by

$$g(t) = [(t-a)/b]^c \tag{1.6}$$

Assuming eqn. 1.6 we obtain the distribution function

$$F(t) = 1 - e^{-[(t-a)/b]^c} \tag{1.7}$$

which was proposed in 1939 (Ref. 1, eqn. (37)). Sometimes, we use $(b-a)$ instead of b, i.e.

$$F(t) = 1 - \exp\left[-\left(\frac{t-a}{b-a}\right)^c\right] \tag{1.8}$$

where t is a variable, a is the minimum-value parameter, b is the characteristic value (63·2 percentile point) and c is the shape parameter.

From the statistical theory of extreme values, a Weibull distribution is also derived. According to Gumbel [2], we consider the stability postulate

$$F^n(x) = F(a_n x + b_n) \tag{1.9}$$

This is a linear transformation which does not change the form of the distribution. As the asymptotic distribution, we consider the distribution of the largest value in samples of size n which are taken from the same population. The two parameters a and b are functions of n. Fisher and Trippett [3] derived the following three asymptotes:

$$F(x) = \exp\{-\exp[-\alpha(x-u)]\} \quad \alpha > 0 \tag{1.10}$$

$$F(x) = \exp[-(v/x)^k] \quad x > 0, v > 0, k < 0 \tag{1.11}$$

$$F(x) = \exp[-(x/v)^k] \quad x < 0, v < 0, k > 0 \tag{1.12}$$

Gnedenko [4] proved that only three asymptotes of eqns. 1.10—1.12 exist under the condition of the stability postulate of eqn. 1.9. Eqn. 1.12 is particularly important for the derivation of the Weibull distribution. This equation was derived from the assumption that $F(0) = 1$ exists; that is the variate is non-positive and the distribution $F(x)$ satisfies $F^n(x) = F(a_n x)$. Here $F^n(x)$ is the distribution of the largest value, where $x = \max(x_1, x_2, \ldots, x_n)$ in n samples $(x_1, x_2, \ldots, x_n)$ from the same population.

Now change the sign of x, that is $y = -x$, and consider n samples $(y_1, y_2, \ldots, y_n) = (-x_1, -x_2, \ldots, -x_n)$ out of the population obeying the distribution $F(y)$. Then the smallest value is written as

$$\begin{aligned} y &= \min(y_1, y_2, \ldots, y_n) \\ &= \min(-x_1, -x_2, \ldots, -x_n) \\ &= -\max(x_1, x_2, \ldots, x_n) \end{aligned} \tag{1.13}$$

If the distribution of the smallest values, $[1 - F(y)]^n$, satisfies

$$F^n(x) = [1 - F(y)]^n = F(a_n x) = 1 - F(a_n y) \tag{1.14}$$

then $F(y)$ is finally written as

$$F(y) = 1 - F(-y) = 1 - \exp[-(-y/v)^k] = 1 - \exp[-(y/b)]^k \tag{1.15}$$

where $b = -v > 0$, $y > 0$, $k > 0$. This is a Weibull distribution.

The yield strength of Bofors steel is a very good example for Weibull distribution [5].

References

1 WEIBULL, W.: 'A statistical theory of strength of materials', IVA-Handl. No. 151, 1939
2 GUMBEL, E. J.: 'Statistics of extremes', (Columbia Univ. Press, 1958)
3 FISHER, R. A. and TIPPET, L. H. C.: 'Limiting forms of the frequency distribution of the largest or smallest member of a sample', *Proc. Cambridge Phil. Soc.*, 1928, **24**, p. 180
4 GNEDENKO, B. V.: 'On the role of the maximal summand in the summation of independent random variables', *Ukarain. Mat. Jl*, 1953, **5**, p. 291
5 WEIBULL, W.: 'A statistical distribution function of wide applicability', *Jl Applied Mechs.*, Sept. 1951, pp. 293–297

Chapter 2

Radar clutter as statistics with Weibull distribution

Radar clutter is the vector sum at the radar antenna of many echo signals from many small scatterers, such as land terrain, clouds, sea waves etc., which are located within the same radar resolution cell. Owing to the movement of these scatterers and the scanning of the radar antenna beam, the amplitude and phase of these echo signals will change, and the amplitude and phase of the vector sum will also change. This is the mechanism of the fluctuation phenomenon of clutter echo. Owing to this phenomenon, the clutter signals become a stationary, and even a non-stationary, random sequence.

If the direction of the radar antenna is fixed, and the clutter data are collected from a fixed range bin, this fluctuation is called temporal fluctuation. If the antenna is rotated, and the clutter data are collected still from the same range bin, this fluctuation is termed spatial fluctuation.

When the antenna is rotating, the type of clutter will be changed even within the same range bin. We will restrict ourselves to the same type of clutter for the spatial fluctuation case.

The amplitude distribution of radar clutter depends not only on the type of clutter but also on the size of resolution cell of the radar and grazing angle of the antenna beam.

For homogeneous clutter, such as desert, clouds, rain, snow and chaff, Rayleigh distributions were reported by many authors. Rayleigh distributions were also observed with low-resolution radars. The larger the size of resolution cell, the more scatterers it contains.

However, as the resolution cell decreases in size, the clutter distribution develops a larger tail than the Rayleigh distribution, especially for sea clutter. In an attempt to remedy this situation, Ballard [1] considered the log-normal description of sea clutter and related the standard deviation of the distribution to the radar's illuminated patch area. Trunk [2] published the data taken by NRL in 1967 at a grazing angle of 4·7° using an X-band radar with vertical polarisation and a 0·02 μs pulse. These data fitted the log-normal curve very well.

The probability density function for the log-normal distribution can be obtained from the normal distribution by using the transformation $X = \ln Y$

$$p(Y) = \frac{1}{Ys\sqrt{2\pi}} \exp\left[-\frac{1}{2s^2}(\ln Y - \mu)^2\right]$$
$$= \frac{1}{Ys\sqrt{2\pi}} \exp\left[-\frac{1}{2s^2}\left(\ln \frac{Y}{Y_m}\right)^2\right] \qquad (2.1)$$

where

$Y =$ log-normally distributed variable

$\mu = \ln Y_m$

$Y_m =$ median value of Y

$s =$ the standard deviation of $\ln(Y/Y_m)$

It is sometimes observed that the distributions of radar cross-section (RCS) when expressed in decibels can be approximated by a normal distribution. Then it is said that the RCS is log-normally distributed.

The mean and median of the log-normal distribution are related by

$$10 \log \bar{\sigma}_0 = 10 \log \sigma_{0m} + 0{\cdot}115\sigma_{0s}^2 \qquad (2.2)$$

where $\bar{\sigma}_0$ is the mean RCS. σ_{0m} is the median RCS, and σ_{0s} is the standard deviation of $10 \log \sigma_0$.

However, the tail of log-normal distribution is sometimes too long to fit practical clutter data, especially for ground clutter.

The Weibull distribution of land clutter was first suggested by Boothe in 1969 [3]. It has a smaller tail than log-normal, and can change the size of the tail by changing the shape parameter.

If we use the backscatter coefficient σ_0 (normalised RCS) of the clutter as the variable instead of the normalised amplitude of the output of detector, the Weibull distribution can be expressed by

$$p(\sigma_0) = \frac{1}{\bar{\sigma}_0} c\sigma_0^{(c-1)} \exp\left(\frac{-\sigma_0^c}{\bar{\sigma}_0}\right) \qquad (2.3)$$

where c is the shape parameter and $\bar{\sigma}_0$ is the average backscatter coefficient. We notice that, for $c = 1$, eqn. (2.3) reduces to the exponential (Rayleigh amplitude) density function.

Eqn. (2.3) is easily integrated to obtain

$$P = \int_0^{\sigma_0} p(\sigma_0)\, d\sigma_0 = 1 - \exp(-\sigma_0^c/\bar{\sigma}_0). \qquad (2.4)$$

Eqn. 2.4 can be solved for σ_0 to obtain

$$\sigma_0 = \{\bar{\sigma}_0 \ln[1/(1-P)]\}^{1/c} \qquad (2.5)$$

The median value σ_{0m} is determined by putting $P = 1/2$ in (2.5), that is,

$$\sigma_{0m} = (\bar{\sigma}_0 \ln 2)^{1/c} \tag{2.6}$$

From eqns. 2.5 and 2.6, we obtain

$$\sigma_0 = \sigma_{0m}/(\ln 2)^{1/c}\{\ln^{1/c}[1/(1-P)]\} \tag{2.7}$$

Taking the logarithm of σ_0 and multiplying by 10, one obtains

$$10 \log \sigma_0 = 10 \log \sigma_{0m} - 10a \log(\ln 2) + 10a \log[\ln 1/(1-P)] \tag{2.8}$$

which is

$$\sigma_0(\mathrm{dB}) = \sigma_{0m}(\mathrm{dB}) + 1{\cdot}6a + 10a \log\{\ln[1/(1-P)]\} \tag{2.9}$$

where

$$a = 1/c$$

If the Weibull probability distribution is plotted on graph paper having the probability scale proportional to $\log\{\ln[1/(1-P)]\}$ and in decibels on a linear scale, the result is a straight line having a slope determined by the parameter $a = 1/c$.

Jakeman and Pusey [4] proposed the K-distribution for the sea-clutter model. This is a class of modified Bessel-function distribution. The probability density function of K-distribution is given by

$$p(x) = \frac{4k}{\Gamma(v)}(kx)^v K_{v-1}(2kx) \tag{2.10}$$

where $K_v(x)$ is a modified Bessel function, k is a scale parameter and v is a shape parameter.

The nth moments of the K-distribution are given by

$$\bar{x}^n = \frac{1}{k^n}\frac{\Gamma(v + n/2)}{\Gamma(v)}(n/2 + 1) \tag{2.11}$$

They used the sea clutter data obtained by Bishop [5] with an X-band radar, and found that it can be fitted with K-distribution very well. However, there is very small difference between K-distribution and Weibull distribution. Fig. 2.1 shows the comparison between Rayleigh, gamma, Weibull, log-normal, and K-distribution with the same second moment equal to 3. The calculation of K-distribution is much more difficult than Weibull distribution. Therefore, we prefer the Weibull distribution rather than the K-distribution as the model of radar clutter.

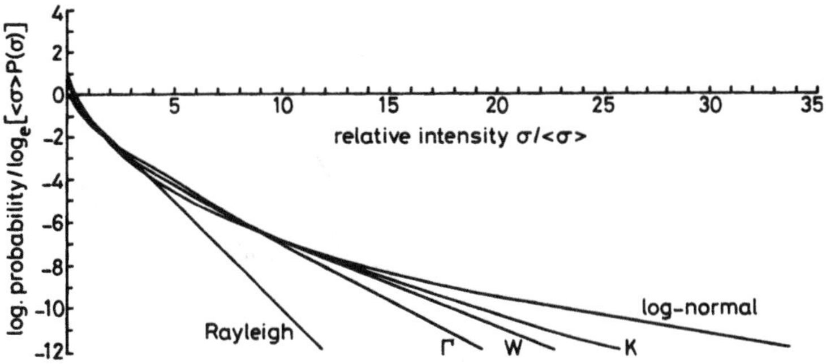

Fig. 2.1 *Comparison of various distributions for second moment equal to 3*

2.1 Land clutter

The use of Weibull distribution to model ground clutter was first suggested by Boothe [3] in 1969. Several sources of measured data of clutter backscatter coefficient have been obtained which exhibit the Weibull distribution. The data points have been plotted on Weibull paper as shown in Figs. 2.2 and 2.3 such that a distribution which obeys Weibull results in a straight line. These sources represent measurements from both high- and low-resolution radars, relatively long and short clutter cell averaging times, and for various frequencies. The greatest deviations from the straight line appear at the very small values, which might have been a result of the receiver noise level.

Ground clutter amplitude from the Rocky Mountains [6—8] obeys a Weibull distribution with a shape parameter of 0·512 for S-band, a 2·0 μs pulsewidth, and a 1·5° beamwidth. The data were taken for a range less than 32 km from an unmasked location. For low rolling wooded hills, grassland, wooded mountains and some man-made structures, taken near Huntsville, Alabama, [9] a Weibull distribution with a shape parameter of 0·626 was observed from L-band radar out to a maximum range of approximately 40 km with a resolution cell defined by a 3·0 μs pulsewidth and a 1·7° beamwidth. The radar site was located on one of the hills such that the depression angle from the antenna, as determined from maps of the area, was generally less than 0·5°. The data taken with a nonscanning antenna and an averaging time of 3 s at each position, represent the strongest 241 clutter cells out of several thousand cells examined, and therefore exhibit a higher median value when compared with other measurements. Even though the smaller clutter cells were neglected, the Weibull distribution still represents a good fit to the distribution of the measured clutter data.

In Sweden experimental studies of the amplitude distribution of radar terrain return have been performed by Linell at the Research Institute of National Defence since the spring of 1961 [10]. The amplitudes of the echoes

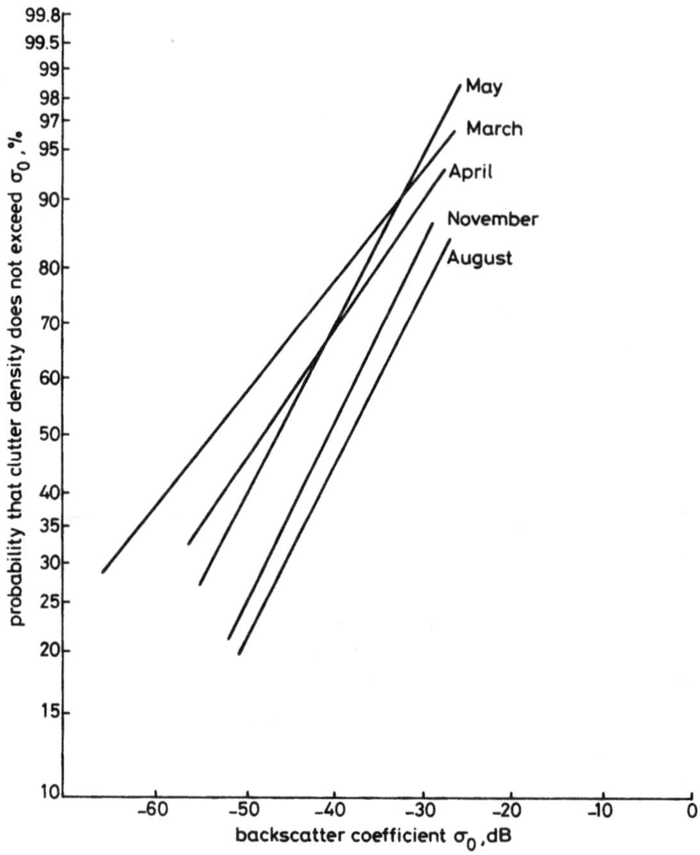

Fig. 2.2 *σ_0 for cultivated terrain at different times of year*
Depression angle 1·25° and resolution 0·17 μs × 1·4°
March (a = 3·82), April (a = 3·3), May (a = 2·84), August (a = 2·83), November (a = 2·6)
The Weibull parameter $c = 1/a$ is determined by slope a of the straight line

returned from the ground, when illuminated by a pulse radar, are measured with the aid of a tower-mounted search radar. The collected data are treated statistically. A 3 cm search radar was placed on top of an approximately 100 ft high waterworks tower. The search antenna has a horizontal lobe width of 1·4° and a vertical lobe width of 30°. When the antenna is fixed, i.e. nonrotating, the illuminated area on the ground has the form of a sector of a circle, 1·4° in width and of infinite radius. During the measurements, the antenna is rotating and a metal finger on it actuates two movable microswitches fastened to the antenna mount. The microswitches in turn act on a relay so that only echoes received when the antenna is pointing within a sector, whose limits are determined by the positions of the microswitches,

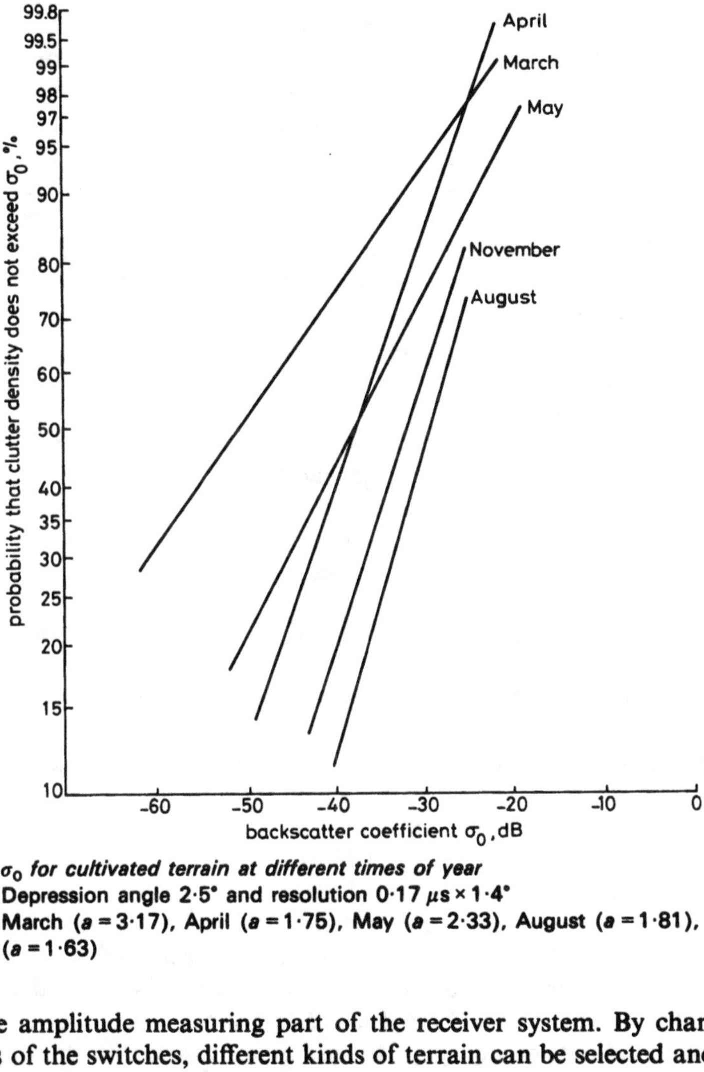

Fig. 2.3 σ_0 *for cultivated terrain at different times of year*
Depression angle 2·5° and resolution 0·17 μs × 1·4°
March (a = 3·17), April (a = 1·75), May (a = 2·33), August (a = 1·81), November (a = 1·63)

reach the amplitude measuring part of the receiver system. By changing the positions of the switches, different kinds of terrain can be selected and studied with respect to their echo amplitudes.

Echoes from the selected sector are amplified and rectified in the radar receiver and fed to the range gate unit. The range gate is opened for 0·17 μs (equivalent to the transmitter pulse duration), corresponding to a geometric length of about 25 m, by a gating pulse, whose position in relation to the transmitter pulse is controlled by two time modulators. In this way only echoes from one 25 m strip of the ground can pass the range gate. The gated signal is amplified and passed to an amplitude sensing device, which has a number of pulse counters on its output. The counters will count the pulses that exceed different amplitude levels in the sensing device.

The measurements can be made up to a maximum range of 12 km. This total distance is divided into a number of coarse ranges, the initial values of which are determined by the manual setting of the delay time of the coarse time modulator. The fine time modulator is triggered by the delayed pulse from the coarse time modulator. The delay time of the fine time modulator is controlled by a stepping relay, whose position determines the value of a resistance in the modulator timing network. By the action of one of the antenna switches, the stepping relay advances one step for each revolution of the antenna. The values of the resistances switched into the timing network are chosen so that the selected coarse range will be successively scanned in range from its inner to its outer limit. The stepping action is automatically stopped when the outer limit is reached, and when this has occurred the amplitude sensing device will be disconnected from the range gate unit. The pulse counters can then be read.

As a suitable place for the experiments a waterworks tower in southern Sweden was chosen. The surroundings of the tower contain many terrain types of interest.

The radar transmitter has a peak power of 40 kW, the pulsewidth is 0·17 μs, the wavelength is 3·2 cm and the antenna rotating speed is 25 rev/min. The antenna is provided with feeding horns for both vertical and horizontal polarisation. The radar receiver has a logarithmic IF amplifier with a dynamic range of 80 dB.

The experimental results are obtained in the form of percentage values, giving the relative number of pulses that have exceeded the 12 different and equally spaced amplitude levels. These percentage values are plotted against amplitude-level number in a Weibull distribution diagram of eqn. 2.9. Figs. 2.2 and 2.3 show results for cultivated terrain at different times of year at X-band with resolution of 0·17 μs × 1·4° and depression angles of 1·25° and 2·5°, respectively.

The main results of Boothe are summarised in Tables 2.1 and 2.2. It can be seen from Table 2.1 that the shape parameter c is decreased with decreasing of the size of resolution cell; but it is not so obvious owing to the different terrains and frequency band. From Table 2.2, it is also easily seen that, as the depression angle (and consequently the median σ_{0m}) increases, the slope of the distribution approaches that of the exponential (or Rayleigh amplitude) distribution. This trend was also evident in measurements of the same terrain at the other seasons of the year. This effect is probably attributed to increased shadowing effects at smaller depression angles.

The fact that the shape parameter c changes with the depression angle can be explained. Possible reasons include: (*a*) the undulating terrain seems rougher with decreasing depression angle; (*b*) the steep cliff of rocky mountain will increase the RCS with the decrease of depression angle; (*c*) echoes from strong reflecting man-made objects, such as buildings, water tanks and fences, tends to be largest for a line of sight near to horizontal; and (*d*) lowlands

Table 2.1 *Shape parameters for different size of resolution cells*

Terrain type	Frequency band	Pulsewidth (μs)	Beamwidth (deg)	Depression angle (deg)	Shape parameter
Wooded hills	L	3·0	1·7	0·5	0·314
Rocky mountains	S	2·0	1·5	—	0·256
Forest	X	0·17	1·4	0·7	0·253–0·266

Table 2.2 *Shape parameter varies with depression angle for a high-resolution ($0{\cdot}17\,\mu s \times 1{\cdot}4°$) X-band radar*

Terrain type	Seasons	Depression angle (deg)	Shape parameter c
Forest	March–May, Aug.	0·7	0·253
Forest	November	0·7	0·266
Cultivated land	April	1·25	0·303
Cultivated land	May	1·25	0·352
Cultivated land	April	2·5	0·573
Cultivated land	May	2·5	0·429
Cultivated land	April	5·0	0·909
Cultivated land	May	5·0	1·0

which are usually flat or gently undulating surfaces are shadowed (masked) by the highlands, e.g. hills or mountains. Therefore, the tail becomes larger (i.e. the shape parameter c becomes smaller) as the depression angle is decreased.

Currie [11] developed a statistical model to describe the effect of depression angle on the shape parameter of Weibull distribution due to the radar antenna scanning over trees at X-band. The results are shown in Fig. 2.4.

Barton [12] also summarised some results of measurements of ground clutter and compared them with the Weibull distribution.

Most of the previous studies of the distribution of ground clutter are limited to spatial distribution. The spatial distribution differs from temporal distribution away to the different mechanisms for the generation of clutter data. For ground clutter, the correlation time is of the order of several hundred milliseconds. So the clutter data, fluctuating with the scanning of the antenna simply reflects the complexity of the terrain being illuminated. If the data is

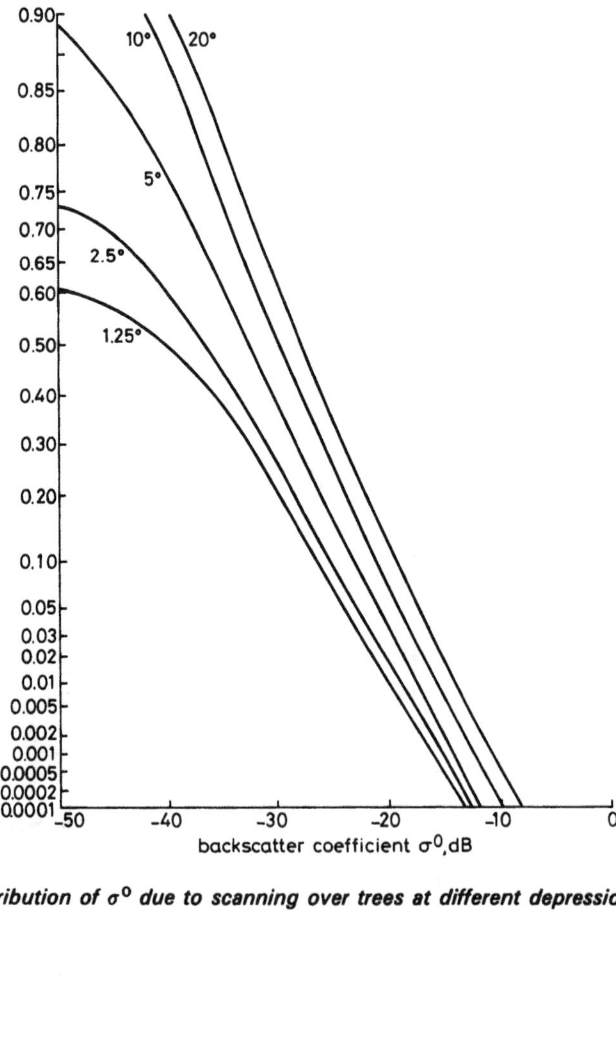

Fig. 2.4 *Distribution of σ^0 due to scanning over trees at different depression angles*

collected from an area with the same type of terrain, the distribution of data is defined as spatial distribution. However, if the data is collected within the same area but with different scans, temporal fluctuation will be introduced. On the other hand, the time series, which is collected from same range bin but a different azimuth sweep, will be a non-stationary random sequence, even if the type of terrain is the same.

Miller [13] studied methods of characterising ground clutter both in spatial and temporal distributions. He suggested a model of the ground clutter consisting of a static spatial distribution of mean values (those of the time series) on which is superimposed a temporal fluctuation whose distribution is that of the time series. Henceforth, it will be assumed that the time series all have the same distribution, but that the mean and variance of this distribution may vary from one point to another.

He studied hill clutter and town clutter with a non-coherent radar (unfortunately, no radar parameter is available). The hill clutter consists of part of the range of the Cotswold Hills to the south of Malvern, England. The town clutter covers approximately the centre of Worcester, which lies to the north-west of Malvern. Parameters describing these recordings are given in the following Table; the spatial spread is the standard deviation of the spatial distribution; the temporal spread is the RMS standard deviation of the temporal distribution.

Type of terrain		Hill clutter	Town clutter
Date		22 March 1982	2 March 1982
Location		Cotswold Hills	Worcester
Noise level (dB)		35	32
No. of	range	51	52
Samples	azimuth	46	25
	scans	60	25
Sampling	range (m)	30	20
Interval	azimuth (deg)	0.15	0.075
	time (s)	15	15
Spatial spread (dB)		8.4	13.1
Temporal spread (dB)		6.3	3.5

The maps of hill clutter and town clutter are shown in Figs. 2.5*a* and *b*, respectively. The spatial distribution and temporal distribution of hill clutter and town clutter are shown in Figs. 2.6*a* and *b*, and Figs. 2.7*a* and *b*, respectively.

The level of the clutter is divided into eight grades in Fig. 2.5. Each grade represents 5 dB in (*a*) and 7·5 dB in (*b*). Darker shading indicates higher return. Noise is present in the lower left-hand corner of Fig. 2.5*a*, and the upper left-hand corner of Fig. 2.5*b*.

The dashed lines in Figs. 2.6 and 2.7 indicate normal distribution, the

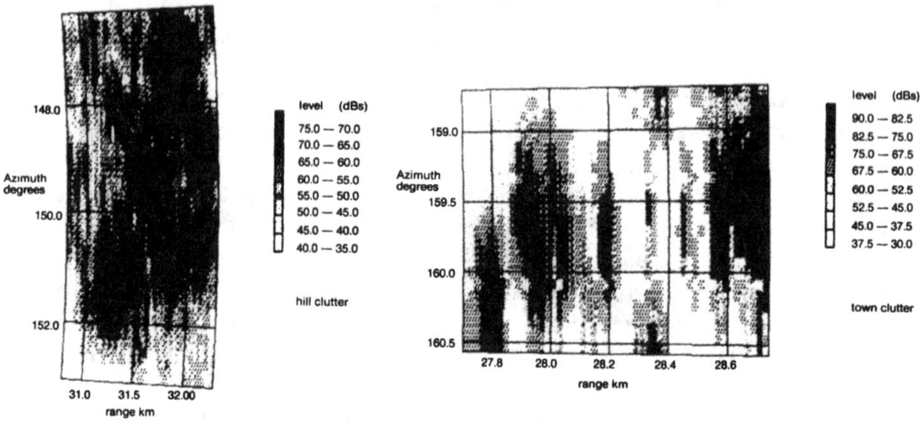

Fig. 2.5 *Maps of hill clutter (a), and town clutter (b)*
Darker shading indicates higher return

dotted lines indicate Weibull distribution, while the solid lines are the results of measurement.

It will be recollected that the spatial distribution was defined as that of the temporal means of the recorded video signal. Since there are many samples available, a reasonable estimate of the distribution may be obtained.

It can be seen from Fig. 2.6 that the spatial distribution deviates from the smooth curve. The distribution for the hill clutter is roughly Weibull, though the tail is more log-normal. For the town clutter, the entire distribution is approximately log-normal, though the fit is not good. Other, larger samples of clutter have shown the tails to be consistently log-normal distributed, with the distribution itself tending to be closer to the Weibull distribution.

The main problem in characterising the temporal distribution of the clutter from the data lies in the fact that the temporal sequences are relatively short. If one had a number of realisations of the time series, a number of estimates could be formed, and the overall estimate could be improved by averaging. However, the time series are not necessarily realisations from the same distribution function, since the parent distributions may have different means and variances. This problem can be overcome by estimating the means and variances for each time series, standardising the time series to zero means and unit variance, and averaging over the estimates of the standardised distribution. The necessity of estimating the means and variances will introduce some additional error in the final estimates of the quantities, but since there seems to be no inherent bias in the technique, the end result should be acceptable.

It can be seen from Fig. 2.7 that, in both examples, the temporal distribution lies approximately midway between the log-normal and the Weibull densities; it is certainly neither Weibull nor log normal. The quantisation due to the small sample size is most noticeable about the tails of the distribution,

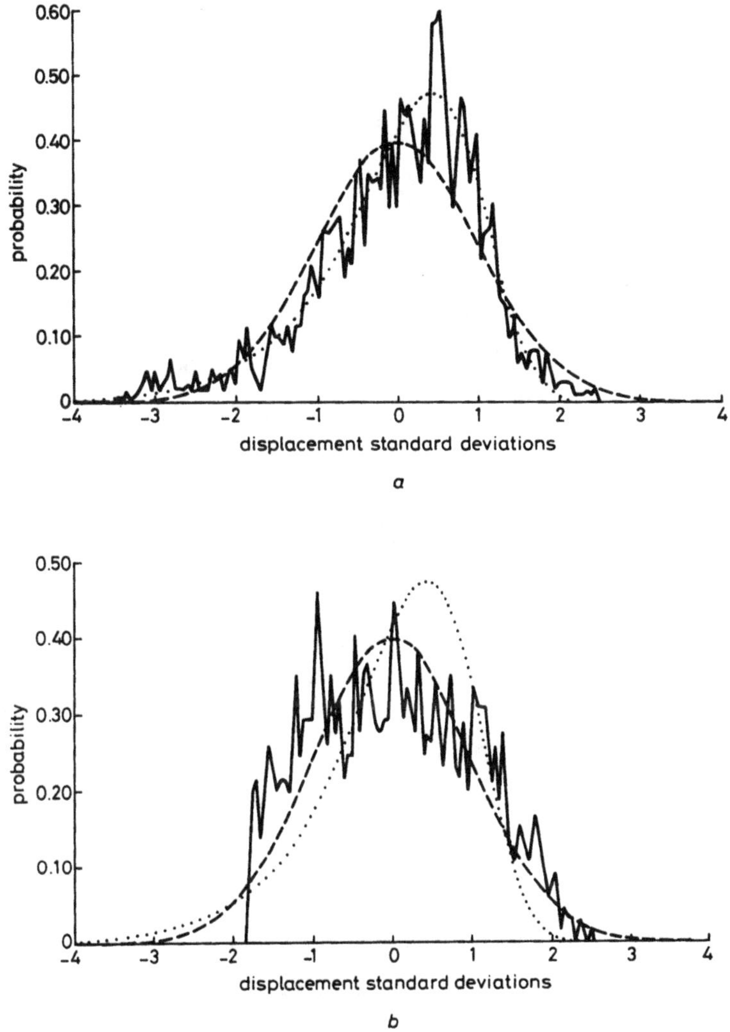

Fig. 2.6 *Spatial distributions for hill clutter (a), and town clutter (b)*

and probably gives a realistic impression of how knowledge of the tails is restricted by the samples size. The form of the distribution is almost certainly broadened by azimuth-pointing errors, since there is no synchronisation between antenna rotation and pulse repetition frequency.

In most of previous studies of amplitude distribution of radar clutter, data were taken from non-coherent radars. In general, the local-oscillator stability of non-coherent radars is not too high in frequency or amplitude. These instabilities will cause either fluctuation in amplitude after mixing and amplitude detection. Furthermore, the dynamic range of the receiver of a

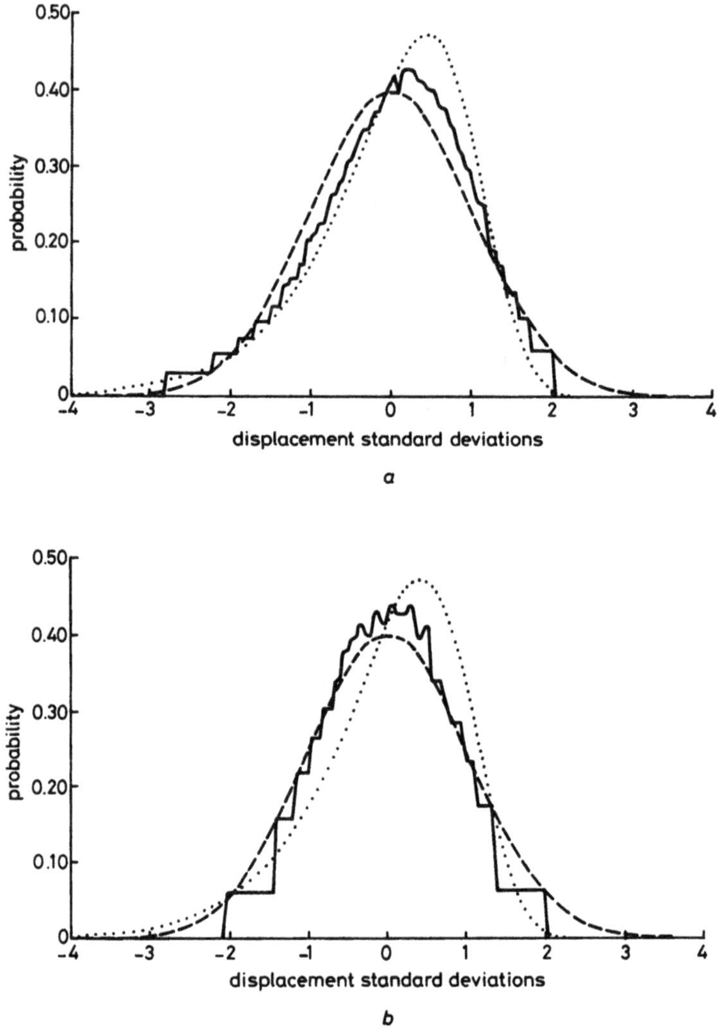

Fig. 2.7 *Temporal distributions for hill clutter (a), and town clutter (b)*

non-coherent radar is also limited by the IF amplifier; clutter echo with large amplitude will be compressed by the IF amplifier. The non-linearity of the amplitude detector will also affect the amplitude distribution, especially in the small-signal region. All these factors may cause inaccuracy of the result of clutter data measurements.

Modern coherent radars or coherent-on-receive radars have highly stable local oscillators and coherent oscillators. The dynamic range of the receiver is large enough for most clutter data measurements, and the linearity of the phase detector is better than that of the amplitude detector. Therefore, measurement of the amplitude distribution of clutter is better carried with a coherent radar or coherent-on-receive radar.

Sekine *et al.* [14], studied the amplitude distribution of ground clutter with a coherent-on-receive L-band ARSR radar. The size of the resolution cell of this radar is 1·23° in azimuth and 3·0 μs in pulsewidth; the type of terrain is cultivated land; the grazing angle is between 0·21° and 0·32°. Data was recorded digitally on magnetic tape, as the inphase component I and quadrature component Q baseband video signals, after passing through an IF amplifier and a phase detector. The sample intervals between two adjacent range bins and between two adjacent range sweeps in the azimuth direction were 0·243 n-miles corresponding to the pulsewidth and 0·1035°, respectively. Each value of I and Q was recorded on the magnetic tape as a 10 bit signal, and hence the minimum and maximum integer values of I and Q were -512 and $+511$, respectively.

To determine the spatial-distribution function of ground clutter with the same type of terrain, a small sample area of 23 range bins, corresponding to 5·6 n-miles, and 20 range sweeps, corresponding to about 2·0° in azimuth, was adopted. The total number of data points in this area is thus equal to 460.

To check whether the amplitude distribution is Weibull, the results of measurements were drawn on Weibull paper. The departure of the measured distribution from Weibull was estimated by calculating the root-mean-square error (RMSE) of the linear fit. The RMSE is the deviation of the data points from the straight line drawn by the least-squares method. The shape parameter of Weibull distribution can also be determined from the slope of this straight line. Since the probability density function as drawn for the amplitude and not for the RCS, the shape parameter $C = 2c$. This means that the shape parameter $C = 2$ for the Rayleigh distribution.

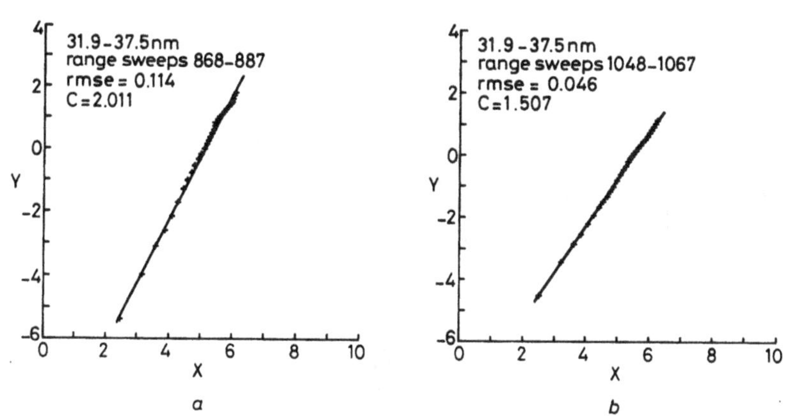

Fig. 2.8 *Examples showing that the ground clutter can be fitted with Weibull distribution but different shape parameters (from Sekine et al. [14], © 1981 IEEE)*

Table 2.3 *Shape parameters for different azimuth sectors but the same type of terrain (from Sekine et al. [14], © 1981 IEEE)*

Azimuth angle (deg)	Sweep numbers	Shape parameter	RMSE
90·57— 92·56	868— 887	2·011	0·114
92·66— 94·64	888— 907	1·843	0·116
94·75— 96·73	908— 927	1·833	0·055
96·83— 98·82	928— 947	1·858	0·069
98·92—100·90	948— 967	1·941	0·125
101·01—102·99	968— 987	1·820	0·102
103·10—105·08	988—1007	1·720	0·063
105·18—107·17	1008—1027	1·841	0·094
107·27—109·25	1028—1047	1·804	0·080
109·36—111·34	1048—1067	1·507	0·046
111·44—113·43	1068—1087	1·670	0·120

The values of shape parameters and RMSE over range sweep numbers 868—1087 in single scan are summarised in Table 2.3.

Two examples are shown in Figs. 2.8*a* and *b*. One unit in X corresponds to 8·7 dB, and one unit in Y corresponds to a probability of 6·6% that clutter will exceed the amplitude given by the abscissa.

It can be seen from Table 2.3 that the distribution of cultivated land can be fitted with Weibull distribution with allowable RMSE, with the shape parameter rated from 1·507 to 2·0.

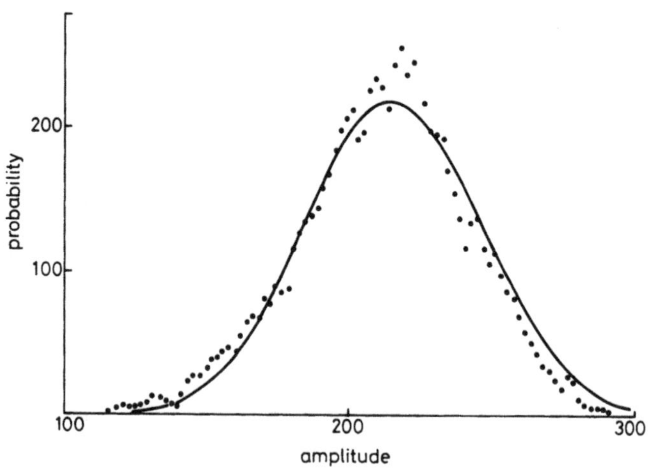

Fig. 2.9 *Temporal distribution of an L-band low-resolution radar in a wooded-hills clutter environment*

Temporal distribution was also measured with same type of radar but with the elevation angle of the antenna beam equal to 2°; the type of terrain is wooded hills. The antenna of the radar was kept stationary, and aimed at the area ranging between 26 and 31·6 n-miles to 224° from the north. The range intervals of 5·6 n-miles were divided into 45 range bins. Therefore, the sample intervals between adjacent range bins were 0·124 n-miles. The result of temporal distribution is shown in Fig. 2.9. The solid line in the Figure is the Gaussian distribution. Therefore, the temporal distribution of this low-resolution radar in a wooded-hills clutter environment is closed to Rician distribution.

2.2 Sea clutter

Sea clutter is somewhat different from land clutter. In general, the backscatter coefficient of sea waves is smaller than that of ground terrain, especially rocky mountains. It increases with grazing angle, radar frequency and sea state; it is greater for vertical than horizontal polarisation and is a maximum upwind, minimum downwind and of intermediate value for cross-wind. In addition, there are often some spikes accompanied by noise-like sea clutter. Since the correlation time of sea clutter is rated from several milliseconds to several tens of milliseconds, 'noise-like' is only in the sense of single sweep or in the range direction. Owing to the long correlation time, sea clutter cannot be removed by non-coherent integration.

One way to reduce the strength of sea clutter is to reduce the resolution cell of the radar. However, as the range cell decreases below about 75 m (0·5 μs), it is observed that the backscattered clutter peaks are approximately constant with respect to time. Only the period between the peaks increases with increased range resolution.

Furthermore, the period between spikes also depends on the polarisation of the radar beam. Fig. 2.10 shows an example of sea spikes for different polarisations. These records were taken with a high-resolution (40 ns × 1°) X-band radar [15]. The radar was installed on a platform site at a water depth of about 30 m and an antenna height of 15 m above mean sea level. These conditions approximated to open sea conditions. The sea state of the record is SS−5 (sea state 5).

Although SS−5 is depicted here, very similar envelopes, albeit with much reduced signal peaks, also occur in SS−2, or even in calm water.

The detail of the clutter spike was observed by Ewell *et al.* [16] with a high-resolution (10 ns × 1·5°) X-band tracking radar at 7° grazing angle when looking in an up/sea direction. A record of this sea spike is shown in Fig. 2.11 for 1800 pulses with 237 Hz repetition rate. A graph of range-cell number versus time indicates that the average velocity of clutter spikes is about 12 knots.

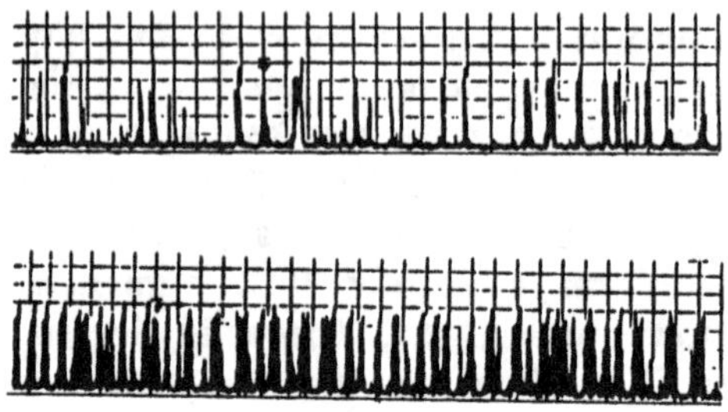

Fig. 2.10 *High-resolution sea spike for different polarisations*
Sea state 5
abscissa: 10 s per division
illuminated cell area: 31·6 m^2
pulse width: 40 ns
top trace: horizontal polarisation
bottom trace: vertical polarisation

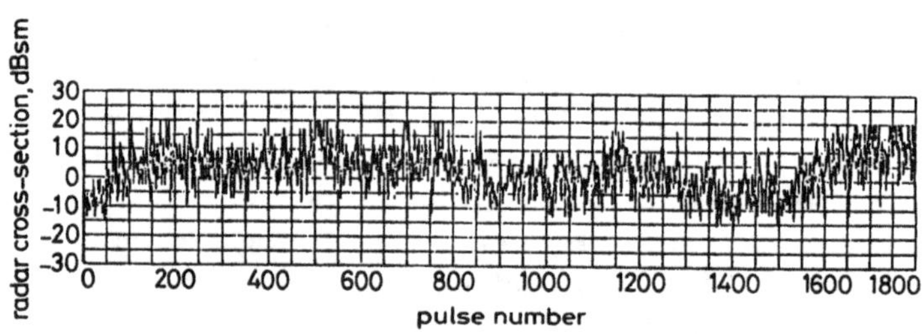

Fig. 2.11 *Clutter spike amplitude time history (from Ewell et al. [16], © 1984 IEEE)*

It can be seen from this Figure that the duration of the clutter spike is at least 6 s, although the correlation time is of the order of several tens of milliseconds.

The average RCS of clutter spikes is at least 10 dB greater than that of the noise-like clutter. Long [17], in a personal communication through Tuley, provded the following comparisons of the complete data run and an analysis of 1600 samples of the spike track:

Statistics	*Complete Run*	*Spike track*
Median normalised RCS	−37·2 dB	−27·7 dB
Mean normalised RCS	−32·9 dB	−24·4 dB
Standard deviation	5·8 dB	4·6 dB

Thus the spikes observed by Ewell *et al.* have rapid fluctuations, similar to extensive-area sea clutter, have magnitudes up to +20 dBsm, appear localised in range and azimuth, and travel at the surface wave speed.

The data of Ewell *et al.* generally seem consistent with the data obtained with lower-resolution X-band radar. Mean RCS was largest for VV polarisation, but the standard deviation of the amplitude fluctuations was largest for HH polarisation.

The long tail in the probability density function was first found in sea-clutter distribution, especially in the high-resolution-radar case. Trunk suggested a log-normal model to fit this long tail. However, sometimes the tail of the log-normal distribution is too long to fit many real clutter data. To solve this problem, the scientists of Johns Hopkins University first suggested the Weibull distribution as the sea-clutter model in an unpublished report [18], which was summarised briefly by Schleher [19].

These data were taken with a relatively high-resolution (0·1 μs × 5·0°) airborne radar for a grazing angle between 1° and 30°. The results were plotted on log-Weibull probability paper. The close fit provided by the straight lines indicates that the data can be closely modeled by Weibull statistics. In addition, the decreasing skewness with increasing grazing angle indicates that the distribution is approaching a Rayleigh distribution. This is shown in Fig. 2.12.

Fay and Clarke [20] also provided an example to show that the log-normal model cannot be fitted well to a set of sea-clutter data, but the Weibull model could be fitted well, especially in the tail part. The data were collected by Bishop with an X-band radar. This radar has selectable pulse lengths of 70 and 270 ns, either vertical or horizontal polarisation, and selectable azimuth beamwidths of 0·6° and 1·2°. The instrumentation is sited on a cliff overlooking the ocean. Slant ranges are quite short, resulting in the azimuth dimension of the clutter patch being about 45 m when using the narrow-azimuth beamwidth. This extremely small clutter patch produces much greater deviation from Rayleigh. However, if these data were fitted with log-normal distribution, the fit deviates in the tail. But the same data can be fitted with Weibull distribution very well. Fig. 2.13 shows an example of a set of sea-clutter data, which were measured in 10—15 knot wind speed, and fitted with log-normal (*a*) and Weibull (*b*) distribution.

In addition to this example, more than 30 sets of data have been evaluated in the same manner with the same conclusion.

As the wind speed increased, the shape parameter of Weibull distribution decreased, and the distribution deviated greatly from Rayleigh distribution. Fig. 2.14 shows the sea-clutter distribution for a wind velocity of 30—40 knots. The parameter a in these Figures is equal to the reciprocal of the shape parameter c, i.e. $a = 1/c$. Therefore, the shape parameter c decreased from 1·24 for 10—15 knot wind velocity to 0·67 for 30—40 knot wind velocity. This means that the distribution has a larger tail in the high sea state.

Weibull-distributed sea-clutter has also been measured with a low-resolution L-band air-route surveillance radar by Sekine *et al.* [21]. The size of the resolution cell is 3·0 μs in range and 1·23° in azimuth. Radar echoes were taken from sea state 3 in a range interval of 23—28·6 n-miles at a fixed azimuth angle of 154°. The direction of the upwind was towards the radar site.

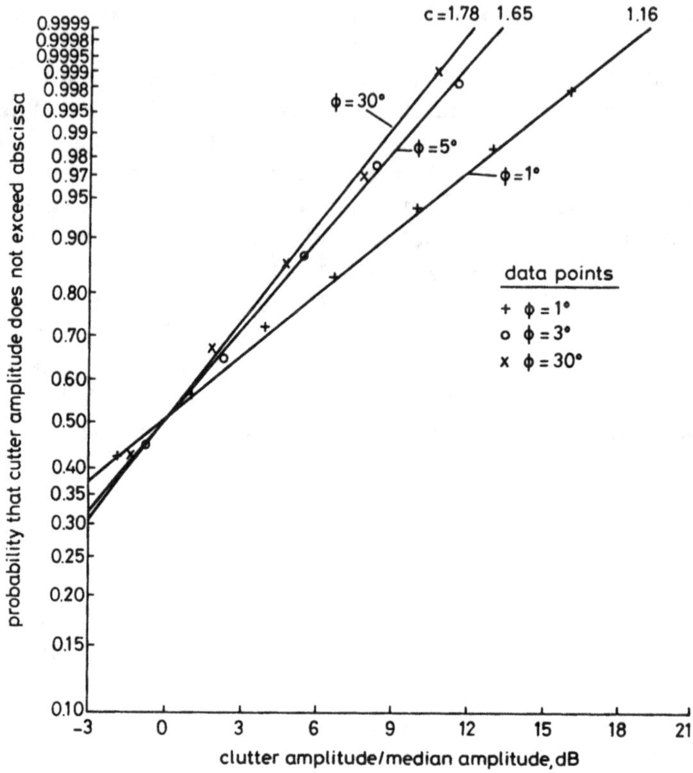

Fig. 2.12 *Sea-clutter data for several grazing angles can be fitted with Weibull distribution (from Schleher [19], © 1976 IEEE)*
Test conditions
Sea state 3
K_u-band
Horizontal polarisation
0·1 μs pulse width
Into sea

The grazing angles were calculated as 0·5° at 28·6 n-miles and 0·72° at 23 n-miles. In this 5·6 n-mile range interval, 23 range bins were sampled. The total number of data points is equal to 120060. Since the antenna was fixed at 154°, these data points were taken from 5220 sweeps. This is equivalent to 15″ according to the 348 Hz repetition frequency.

The result is shown in Fig. 2.15. The departure of the empirical distribution from the Weibull distribution was estimated by calculating the root-mean-square error (RMSE) of the linear fit. The RMSE is the deviation of the data points from the straight line estimated by the least-squares method. From this it can be seen that sea clutter can be fitted with Weibull distribution with a shape parameter of $c = 1{\cdot}585$ with a RMSE $= 0{\cdot}043$.

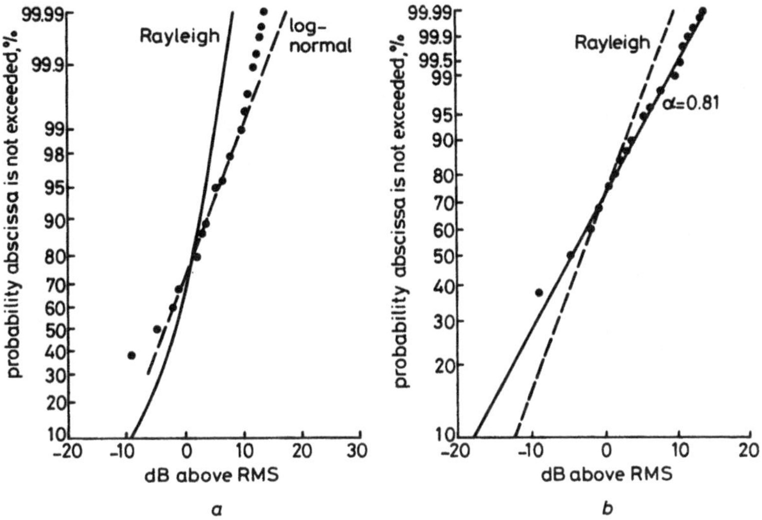

Fig. 2.13 *X-band sea-clutter data for 10—15 knot wind velocity fitted with log-normal* (*a*) *and Weibull* (*b*) *distribution*

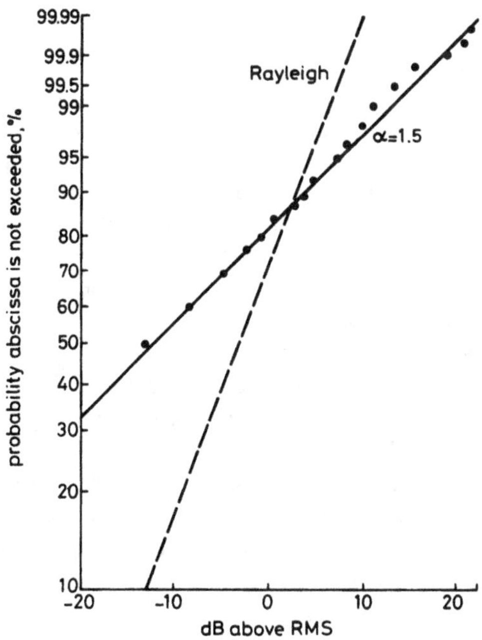

Fig. 2.14 *Weibull-distributed sea-clutter data for 30—40 knot wind velocity*

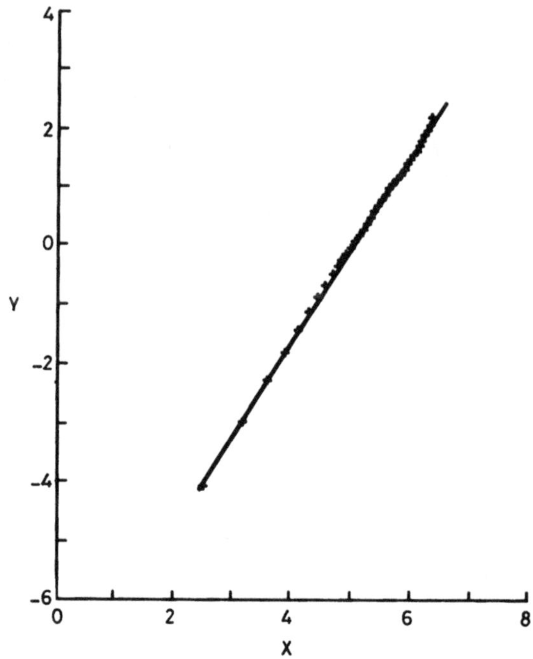

Fig. 2.15 *Sea-clutter distribution of a low-resolution radar fitted with Weibull distribution*

Since the antenna was fixed at 154°, this amplitude distribution belongs to temporal distribution. However, since the range interval of the sample area is equal to 23 range bins, this corresponds a spatial extent of 5·6 n-miles in range direction. Therefore, it is a mix of spatial and temporal distribution.

Sea clutter is different from land clutter in the type of clutter, since there is only the difference in sea state rather than types of terrain. At the time of measurement, there is only one sea state. Therefore, sea clutter is more homogeneous than the land clutter, especially for the low-resolution radars. The temporal distribution of adjacent range bins in the same time period is the same, so the data collected from adjacent range bins of small range interval can be processed together.

Weibull distribution was also obtained with a low-resolution S-band radar at shallow grazing angles even when the arc subtended by the azimuth beamwidth exceeded 4 km [22].

Obviously, the radar designer who desires to use Weibull distribution must have suitable values for the shape parameter c. However, the parameters of the instrumentation that obtained the existing data may not correspond to those of the radar being considered. If the clutter-patch dimensions of the instrumentation are known, one can calculate values of c for larger or smaller clutter patches. This method was proposed by Fay *et al.* [20], and is based on the change in Weibull statistics for doubling clutter patch size; it is derived by considering all possible pairs of clutter samples added at all possible phase angles. The changes in skewness for other clutter-patch sizes are then derived by graphical interpolation.

The clutter from a resolution cell is assumed to be the sum of clutter from two equal halves of that cell. If v_1 and v_2 are the clutter amplitude of two halves, the vector addition gives the combined amplitude v for the total cell. A particular value of v may be made up from many combinations of different v_1, v_2 and phase angles. If $p(v_1)$ and $p(v_2)$ are Weibull distributed, the combined $p(v)$ is also Weibull distributed.

The results are plotted in Fig. 2.16*a* on a linear scale of c. The straight-line relationship of initial c_1 to combined cell c_2 is

$$8{\cdot}5c_2 = 7c_1 + 3 \tag{2.12}$$

This equation has been used to plot points on Fig. 2.16*b* to show the change in Weibull parameter for $\times 2$, $\times 4$, $\times 8$, in patch size and also for $\times 1/2$, $\times 1/4$, $\times 1/8$. These points lie on smooth curves drawn so that Fig. 2.25*b* can be used to find any change in Weibull parameter c for resolution cell changes from $\times 0{\cdot}1$ up to $\times 10$.

Some authors have reported that the amplitude distributions for different polarisations are different. Olin [15] analysed the data collected by Hansen and Cavaleri of NRL, and obtained the result that the amplitude distribution for vertical polarisation can be fitted with Weibull distribution, while the amplitude distribution for horizontal polarisation can be fitted with log-normal distribution. The results are shown in Fig. 2.17*a* and *b*. These data were measured with an X-band radar using a pulse width of 40 ns and a pulse repetition frequency of 2000 Hz. The conditions of the installation of this radar are exactly the same as those of Fig. 2.10.

These Figures are drawn on Weibull or Rayleigh paper. The straight line in these curves represents Weibull distribution. It can be seen that the amplitude distribution for vertical polarisation can be fitted with Weibull distribution very well. However, the amplitude distribution for horizontal polarisation deviates

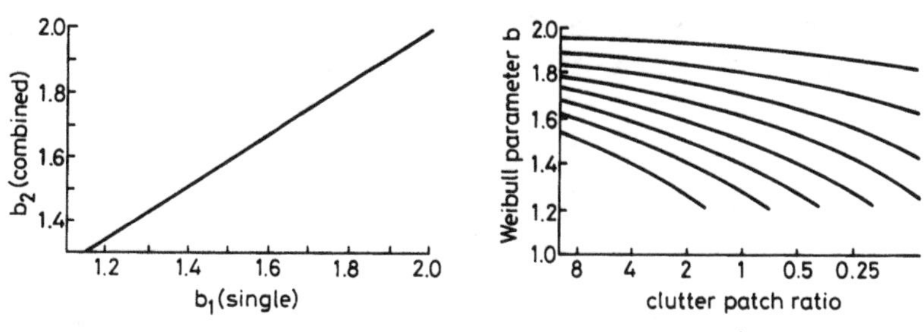

Fig. 2.16 *Change in Weibull parameter with cell size*

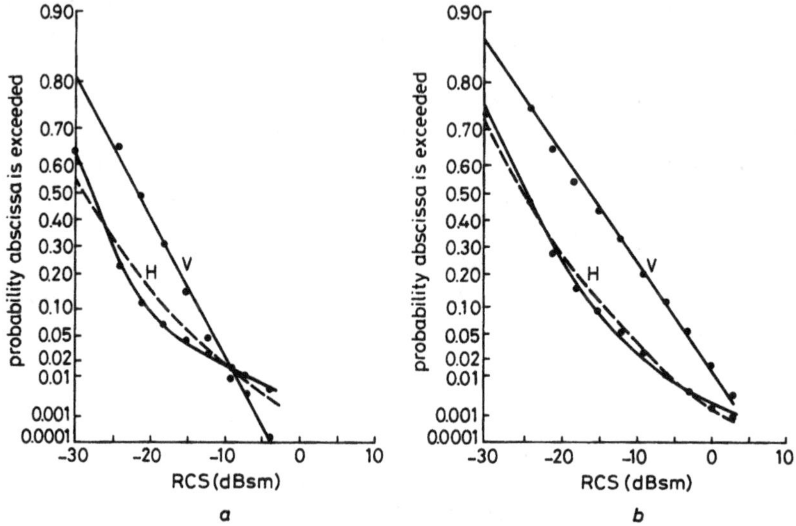

Fig. 2.17 *Amplitude distributions for different polarisations and different sea state*
(a) Sea state 2
(b) Sea state 5

from Weibull distribution, and can be fitted with log-normal distribution as shown with dashed line in these Figures, especially for high sea state.

In these Figures, the abscissa scale is in terms of the actual measured radar cross-section (RCS). The normalised RCS can be obtained by subtracting 15 dB (relative area resolution cell) from these values.

The solid straight line for vertical polarisation in these Figures was fitted by eye and qualitatively estimated to fit the data points. Based on these lines the shape parameter c of Weibull distribution can be determined and is shown in Table 2.4.

It has been suggested [34] that the horizontally polarised component may have two independent sources of fluctuation. One possibility is to model the data using two independent Weibull distributions, the variates of which are simply summed. Thus, given independent Weibull variates σ_{01}, σ_{02} with respective distributions defined by $(\bar{\sigma}_{01}, c_1)$ and $(\bar{\sigma}_{02}, c_2)$, a combined variate $\sigma_0 = (\sigma_{01} + \sigma_{02})$ is formed and the resulting distribution function plotted.

Table 2.4 *Vertical-polarisation Weibull parameters*

Sea state	$\bar{\sigma}_0$ (dB)	Shape parameter c	σ_m (dB > 1 m^2)
SS–2	−11·87	0·622	−21·7
SS–5	−6·42	0·495	−16·2

Using a computer simulation with 10 000 samples for each distribution, fits to the horizontally polarised data were explored. Those shown by the solid curve in Figs. 2.17*a* and *b* represent the best fit qualitatively estimated. Parameters used are given in Table 2.5. The fit to the experimental data is good. It is not surprising, for example, that, with the four independent parameters available, a good fit is obtained.

From this result, it can be seen that the summation of two Weibull-distributed variates with different shape parameters is no longer Weibull distributed. The shape of the tail part of the distribution can be adjusted by adjusting the shape parameters and the mean values of these variates. However, the fitting process is too complex, so it is better to fit these data with log-normal distribution.

Maaloe [23] measured sea clutter with a median-resolution X-band radar. The polarisation of this radar is horizontal. The essential radar characteristics as regards the sea clutter analysis are as follows:

Aerial

Height:	15 m above mean sea level
Polarisation:	Horizontal
Azimuth resolution:	0·8° at 3 dB

Transmitter

PRF:	850 Hz, 1650 Hz, 3300 Hz
Pulse width:	1 μs, 0·25 μs

Receiver

Type:	Superhet—log IF
Bandwidth:	5 MHz or 16 MHz

The bandwidth of the video recorder is 5 MHz. The signals are digitised into 7 bits corresponding to a dynamic range of 42 dB.

The sea-clutter measurement is carried out for ranges of 0·5 to 3·5 n-miles. The weather was dry and the wind speed was about 11 knots, corresponding to sea state 3.

The distributions applied to the sea-clutter amplitudes show that at close ranges, i.e. in a heavy clutter environment, the Weibull distribution appears to be the best fit. At ranges where the clutter dissolves into so-called clutter islands, the clutter amplitudes tend to be better fitted by log-normal distribution. At even larger ranges, where clutter decreases and finally disappears, the amplitude distribution again approaches a Weibull distribution, owing to the fact that this is the distribution of the receiver noise. As an example, Fig. 2.18 shows a histogram which is best fitted with log-normal distribution.

Sekine *et al.* [24] measured sea clutter with a low-resolution L-band ARSR radar with horizontal polarisation. The resolution of the radar is 3 μs × 1·2°. The pulse repetition rate is 350 Hz, and the antenna scan rate is 6 rev/min. Sea echoes were observed in a range interval of 16—21·6 n-miles over an azimuth interval of 128·4° to 172·2° (range sweep numbers 1230—1649). The grazing

Table 2.5 *'Double' Weibull parameters for horizontal polarisation*

Sea state	$\bar{\sigma}_{01}$ (dB)	c_1	σ_{0m1} (dB)	$\bar{\sigma}_{02}$ (dB)	c_2	σ_{0m2} (dB)
SS–2	−22·22	0·833	−29	−7·88	0·172	−55
SS–5	−14·69	0·625	−26	−7·85	0·189	−50

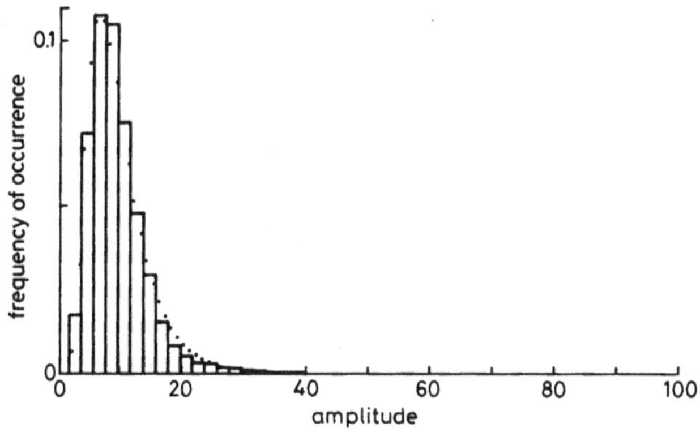

Fig. 2.18 *Histogram of spatial clutter fitted with log-normal distribution*
Range: 2 n-miles
Pulse width: 250 nsec
Wind direction: up-wind
Log-normal fit

angles are calculated as 0·25° at 16 n-miles and 0·13° at 21·6 n-miles. The observed upwind velocity was 17 knots, which is equivalent to sea state 4, and the direction of the wind was south-east.

In-phase and quadrature base-band video signals from six scans were recorded in 10 bits on magnetic tape. The range interval of 5·6 n-miles was divided into 23 range bins, each corresponding to one pulse width. The azimuth interval of 2° corresponds to 20 sweeps. The variation of sea-clutter amplitude was thus examined within a 23×20 sampling area over a 60 s period. The total number of samples is equal to $460 \times$ no. of scans. One example of the first scan for range number 1230—1249 is shown in Fig. 2.19.

It can be seen that the fit to a log-normal (dashed line) is poorest in the tail and a log-Weibull distribution (solid line) determined by the least-squares method is the better fit. Fig. 2.28 shows that the log-normal and log-Weibull distributions appear as normal and Weibull, respectively, when the clutter amplitudes z are plotted against $\ln z$. The log-Weibull distribution shape parameter was calculated as follows:

The log-Weibull distribution function for variable z is generally written as

$$p(x) = \begin{cases} \dfrac{c}{a}\left(\dfrac{x}{a}\right)^{c-1} \exp\left[-\left(\dfrac{x}{a}\right)^{c}\right] & \text{for } x > 0, a > 0 \text{ and } c > 0 \\ 0 & \text{otherwise} \end{cases} \tag{2.13}$$

where $x = \ln z$, a is a scale parameter and c is a shape parameter. Eqn. 2.8 is valid for $z > 1$. The mean value of the receiver noise level corresponds to approximately $z = 10$; therefore very small values of z are insignificant.

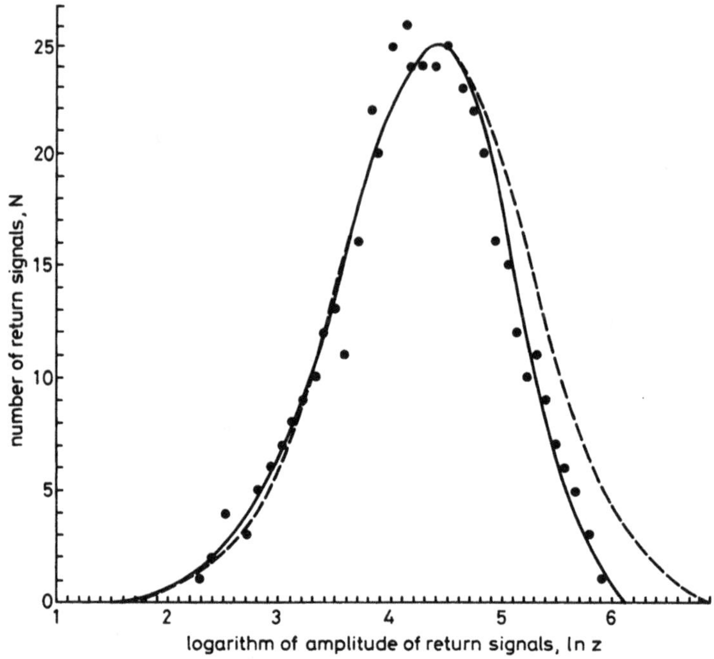

Fig. 2.19 *Number of return signals versus logarithm of amplitude of sea clutter*

Eqn. 2.8 may be rewritten as

$$Y = cX - c \ln b \tag{2.14}$$

where

$$Y = \ln\left\{\ln 1 - \ln\left[1 - \int_0^x p(x)\, dx\right]\right\} \tag{2.15}$$

with $X = \ln x$. From eqn. 2.14 it can be seen that c may be directly estimated from a plot of Y against X.

Six examples, one from each of the six scans for range-sweep number 1230—1249 are shown in Fig. 2.20. A straight line was fitted to the values of Y and X by the least-squares method. The deviation from the straight line for small values of X is due to the receiver noise. The values of c for six scans of 21 sectors of 2° azimuth extent are given in Table 2.6.

Trizna [25] of NRL carried out an open-ocean sea-clutter measurement with a high-resolution shipboard X-band radar. The radar used was a Raytheon Mariner Pathfinder, with a 1° azimuthal-beamwidth antenna with approximately 28° beamwidth in elevation, and a nominal 65 ns pulse width. The polarisation of the antenna is horizontal. The receiver has a log-linear

Table 2.6 *Shape parameters for different azimuth sectors and scans*

	c					
Azimuth (deg)	Scan 1	Scan 2	Scan 3	Scan 4	Scan 5	Scan 6
128·4—130·4	6·00	6·11	6·20	6·18	6·14	6·11
130·5—132·5	6·29	5·82	6·17	6·11	6·34	5·89
132·6—134·6	5·97	6·21	5·87	6·43	5·22	5·73
134·7—136·7	5·98	5·46	5·33	6·38	6·54	6·07
136·8—138·8	6·47	6·12	5·94	6·64	6·58	7·18
138·9—140·9	6·65	6·93	5·80	6·71	5·69	5·53
141·0—143·0	6·41	6·08	5·34	5·80	7·35	6·20
143·1—145·1	5·29	5·94	6·03	5·69	6·32	6·43
145·2—147·2	6·57	6·18	5·91	5·82	5·51	6·07
147·3—149·3	6·24	5·63	5·76	6·07	5·53	5·84
149·3—151·3	6·11	6·33	6·00	6·25	6·32	6·12
151·4—153·4	6·03	6·09	6·32	5·73	6·14	6·41
153·5—155·5	6·53	6·15	5·69	5·72	6·00	5·61
155·6—157·6	5·97	6·42	6·60	6·34	7·35	6·55
157·7—159·7	6·23	6·09	6·50	6·08	6·93	5·93
159·8—161·8	5·76	6·19	6·31	5·78	6·18	5·95
161·9—163·9	6·21	6·15	5·89	6·19	5·59	6·13
164·0—166·0	6·48	5·99	6·43	5·99	6·07	6·51
166·1—168·0	6·19	7·21	6·20	6·92	7·32	6·93
168·1—170·1	6·20	6·20	6·72	6·44	6·51	6·32
170·2—172·2	6·72	6·51	6·34	6·21	6·42	6·25

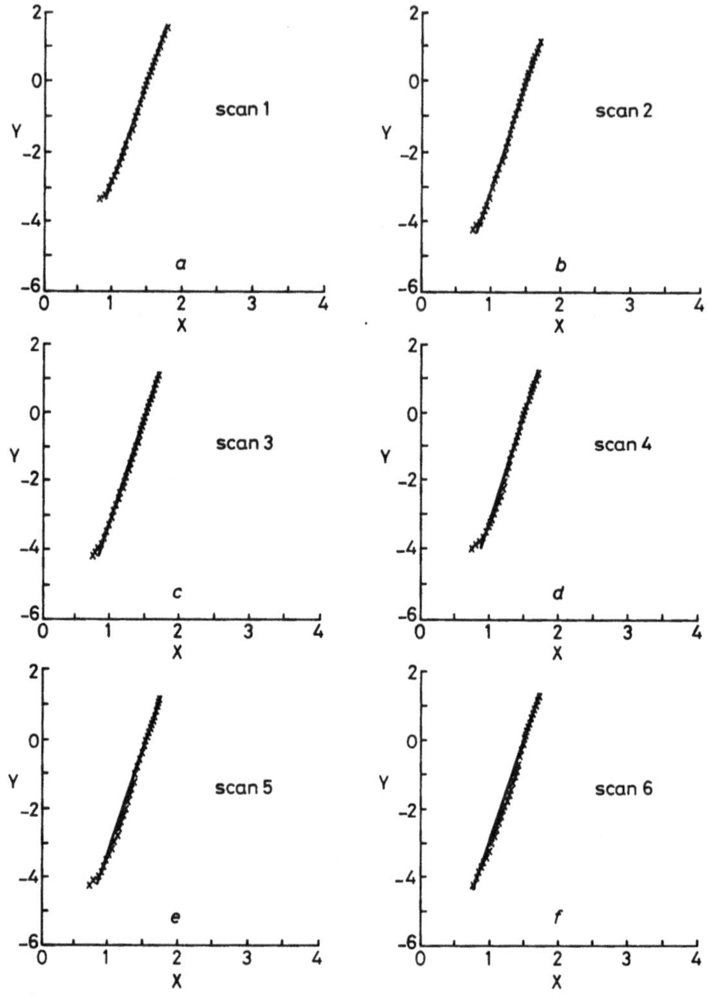

Fig. 2.20 *Sea-clutter data fitted with log-Weibull distribution*

response over nearly 60 dB, and was calibrated along with the digital recording system using a signal generator.

Data were collected over 360° in azimuth and low depression angle, for a variety of wind speeds.

A sample of the recorded output signal from the receiver for one range bin but different azimuth is shown in Fig. 2.21. This Figure is expressed as the relative RCS of the sea clutter with the angle to wind. The depression angle is 1·03° and the wind direction is downwind. This plot gives an indication of the range of clutter amplitudes to expect from a single range sample, and shows its variation with azimuth.

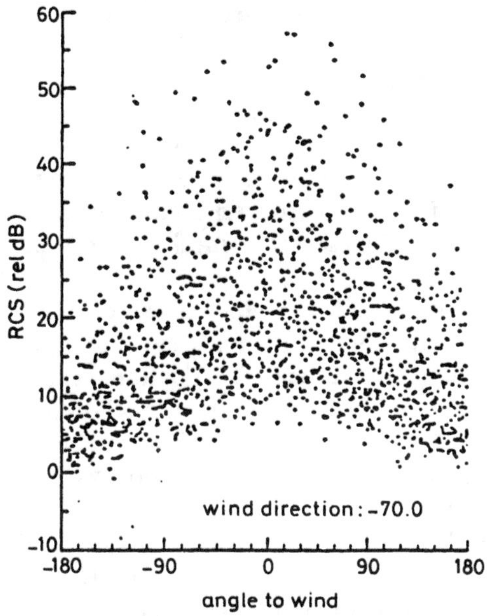

Fig. 2.21 *Relative RCS versus angle to wind (from Trizna [25], © 1985 IEEE)*

When illuminating in the upwind direction, the sea clutter is found to be roughly periodic with the spacing of the dominant ocean wavelength. An example of this behaviour is shown in Fig. 2.22, in which amplitude is shown on an arbitrary decibel scale, and the range ticks are in 30 m units. The radar signal showed this characteristic over all depression angles processed (1°—7°), indicating that the strongest scattering centres at these low angles are probably associated with the peaks of waves. These waves are generated by winds of about 15 knots and higher. The dominant wave period is about 4·4 s for 30 m ocean wavelength.

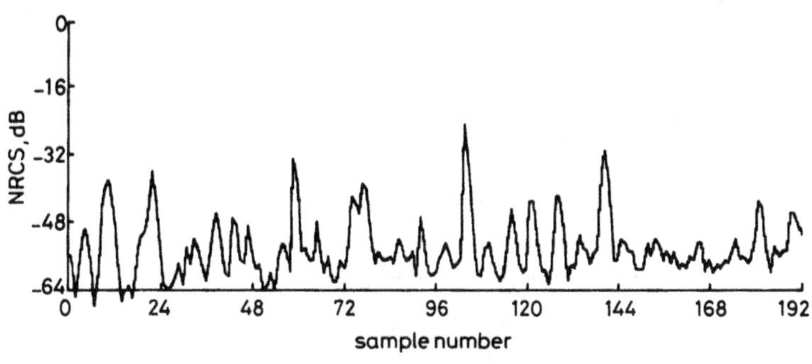

Fig. 2.22 *Normalised RCS versus range for upwind illumination*
Range units in sample number, at 4·5 m per sample (from Trizna [25], © 1985 IEEE)

The cumulative distribution of the sea clutter was calculated and is shown in Fig. 2.23.

These data were selected from a 40° azimuthal sector around the wind direction, corrected for the receiver gain characteristics, and transformed to an accurate representation of the received power. The results were then converted to normalised RCS, or σ_0. The cumulative distribution is plotted on Rayleigh paper. Three distributions are easily separable; linear curves with different slopes representing two of these are identifiable as Weibull distributions on such paper.

The leftmost set of samples for the lowest values of NRCS in Fig. 2.23 are simply noise samples. The slope of the mid-region straight line is just the same as those of Olin's plots for vertical polarisation. The values of NRCS for horizontal polarisation in the mid-region are 4—6 dB lower than that of vertical polarisation. This is the order of difference one can expect for the level of Bragg scatter from small-scale roughness with wavelengths of one-half the radar wavelength. Hence, one can tentatively identify the scattering machinism responsible for this mid-region in the plot of Fig. 2.32 as being due to Bragg scatter from the portions of the wave illuminated away from the crests.

The sea-spike phenomena for horizontal polarisation could be responsible for the linear fit in the highest-amplitude region of the cumulative distribution.

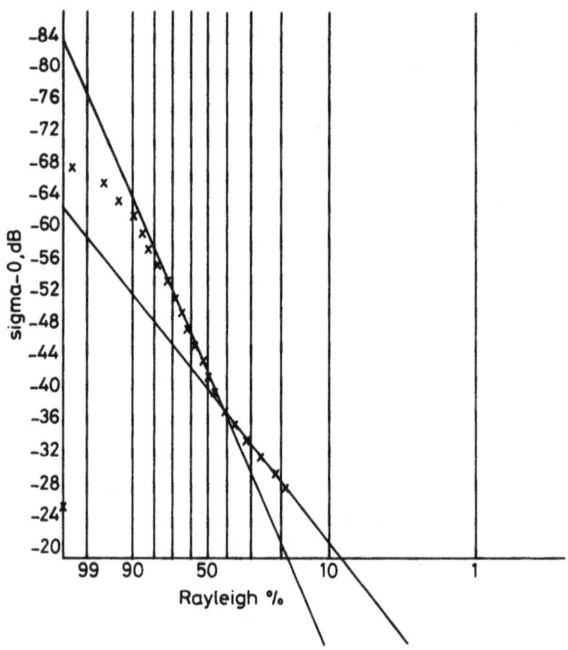

Fig. 2.23 *Cumulative distribution of normalised RCS of sea clutter (from Trizna [25], © 1985 IEEE)*

'Sea spikes' or 'bursts' have been reported by Kalmykov and Pustovoytenko [26], Kalmykov *et al.* [27], and Lewis and Olin [28], for horizontal polarisation, using high spatial resolution. Their defining characteristic is the large amplitude and long lifetime when compared to the highly decorrelated return presented by the remaining time periods in a time series of radar returns. This decorrelated radar return becomes less prominent as the pulse is shortened—as one might expect for a localised scatter in the presence of spatially distributed clutter, as presented by the Bragg scatter mechanism. Breaking-wave features have been associated with these returns in the Russian literature, but Lewis and Olin have reported that nonbreaking wave crests are sufficient to produce such scatter.

Slopes were determined from cumulative distributions similar to Fig. 2.23, which are plotted on Rayleigh-probability paper. On such a co-ordinate system, a Rayleigh distribution would have a unit slope (−1 for RCS increasing downward as presented here). The slopes of mid-region straight lines are called Bragg slope, which is changed with the wind speed. A set of Bragg-slope magnitudes are shown plotted in Fig. 2.24 against wind speed.

Two sets of data for a similar depression angle processed by Olin in Table 2.4 showed slopes of 1·61 and 2·02 for local 'sea state' values of 2 and 5, respectively, which cover wind speeds between 7—15 knots and 27—40 knots. These values fit within Fig. 2.24, using a scaling of 10 m/s equal to 20 knots.

One can hypothesise that the distribution is a combination of two Weibull distributions, such that the first is assumed to hold below some cutoff RCS value, and the second holds above the cutoff. Physically, this means that either

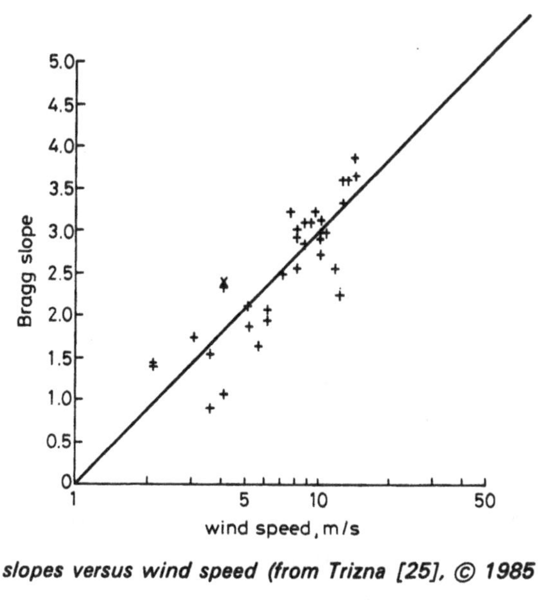

Fig. 2.24 *Bragg slopes versus wind speed (from Trizna [25], © 1985 IEEE)*

one scattering mechanism is present or the other, but not both, either side of some cutoff RCS value. However, one can still parameterise each distribution independently for fitting purposes, as if one or the other were only present. If this hypothesis is accepted, it is useful to consider the intersection of each Weibull linear fit extended through the 50% point, and identify these as medians of each independent distribution. The result of doing this is shown in Fig. 2.25 for the Bragg effects, plotted versus wind speed. This Figure shows the relationship between Bragg medians and wind speed. The data are seen to cluster stronger for higher wind speeds, perhaps because the effects of air–sea temperature difference are stronger at the lower wind speeds, thus causing more scatter of data in this region.

Fig. 2.26 shows sea-spike medians versus wind speed for a 1·26° depression angle, and contains the largest normalised RCS in the 10 m/s wind-speed region of all depression angles that were processed. The 50% intercept is interpreted, as before, as a normalised RCS median at X-band, but should really be translated to RCS if the scatterer responsible for the sea spikes is indeed a localised wave crest which may not fill the beam. For the pulse length (65 ns) and beamwidth (1°) used, a −30 dB NRCS gives −7·3 dB m^2 or 0·19 s m^2 total cross-section. However, if the scatterer were localised in the range dimension to the order of 1 m, the true median NRCS would be of the order of −20 dB. For a range bin some 350 m further in range, the median cross-section was about 1 dB smaller in magnitude, probably due to the scatterer not filling the beam in the transverse dimension.

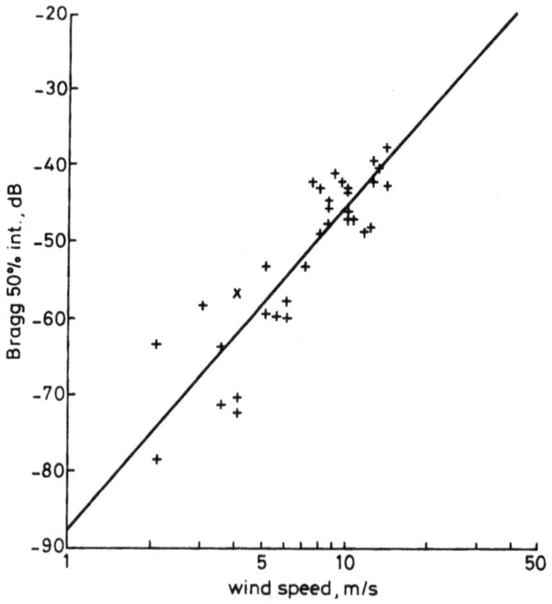

Fig. 2.25 *Bragg medians versus wind speed (from Trizna [25], © 1985 IEEE)*

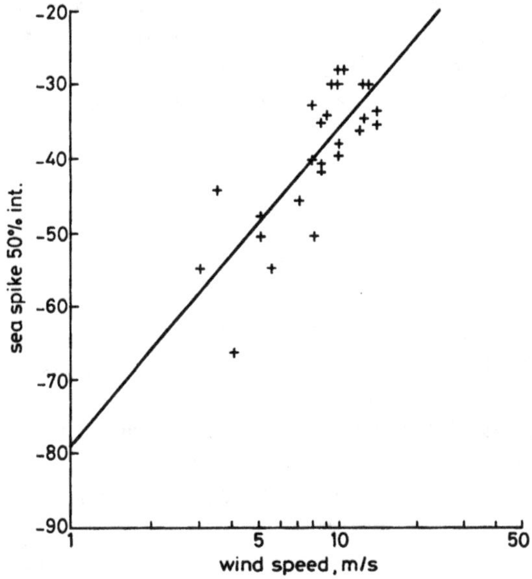

Fig. 2.26 *Sea-spike medians versus wind speed (from Trizna [25], © 1985 IEEE)*

Another useful parameter to plot is the percentage co-ordinate of the intersection of the two linear fits of Fig. 2.23. If one accepts the model for the two scattering mechanisms, this plot can be interpreted as a rough indication of the percentage of RCS values which are sea-spike returns. Fig. 2.27 shows a plot of this parameter versus wind speed.

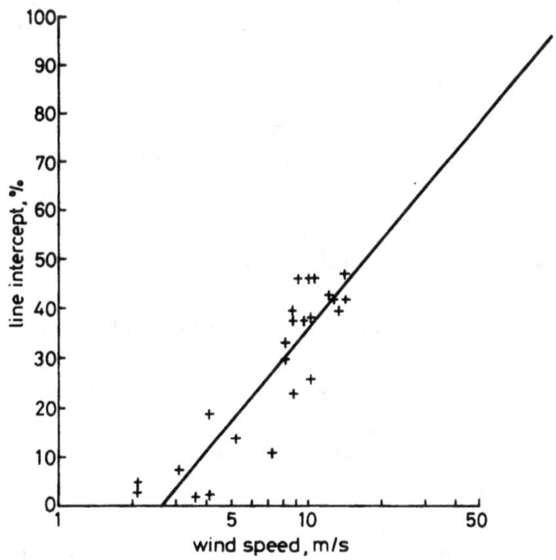

Fig. 2.27 *Intercept of two linear fits versus wind speed (from Trizna [25], © 1985 IEEE)*

From Fig. 2.26 it appears that the sea-spike median cross-sections dominate the Bragg medians over the entire range of wind speeds shown. However, these results must be considered in conjunction with Fig. 2.27 of the percentage of total returns which were sea-spikes. This percentage tends to decrease for small depression angles, which is probably a result of shadowing effects.

2.3 Sea-ice clutter

The backscatter coefficient of sea ice is much greater than that of sea waves. It is well known that the backscatter coefficient of sea clutter mainly depends on the sea state, while the backscatter coefficient of sea ice mainly depends on the type and thickness of the sea ice. In general, sea ice includes fast ice and pack ice. Fast ice is defined as stationary ice near the coast, while pack ice is a moving ice. Pack ice collides together and hummocks are formed, which grow from a height of a few meters to tens of meters.

Radar returns from sea ice depend not only on the volumetric structure, but also on the surface roughness and dielectric constant. These, in turn, depend on the age, thickness, development history and temperature of the ice.

The formation of sea ice is complex and depends on the brine content of the sea surface water, temperature, vertical salinity profile, and depth of water. Foir sea ice to form, sea water, because of its salinity, must be cooled to temperatures below the freezing point of fresh water. Needle-like fragile ice, containing spherical ice crystals, is formed when sea water is cooled below its freezing temperature. With further cooling these crystals grow rapidly and close together to form a uniform sheet of ice known as young ice.

In its first year ice grows to a thickness of more than a meter. During this year it is classified as thin (<30 cm) and thick (>30 cm) first-year ice. The ice surface melts during summer and refreezes during winter, and the thickness also increases further over the years. Ice that has undergone several melt-and-refreeze cycles and has a thickness of more than 2 m is called multiyear ice. The salinity of young ice is much higher than that of first-year ice, and the salinity of first-year ice is much higher than that of multiyear ice. Therefore, the backscatter coefficient of multiyear ice is greater than that of fresh-water (lake) ice, and the backscatter coefficient of first-year ice is greater than that of multiyear ice.

The backscatter coefficients of different types of sea ice versus incident angles are shown in Fig. 2.28 [29]. it should be noted that the abscissa of this Figure is the incident angle, not grazing angle or depression angle. It can be seen that the backscatter coefficients of sea ice at 80° incident angle are about -32 dB. This Figure corresponds to that of sea clutter in sea states 3 to 4 at 10° grazing angle. It is about 10 dB greater than that of sea clutter in sea state

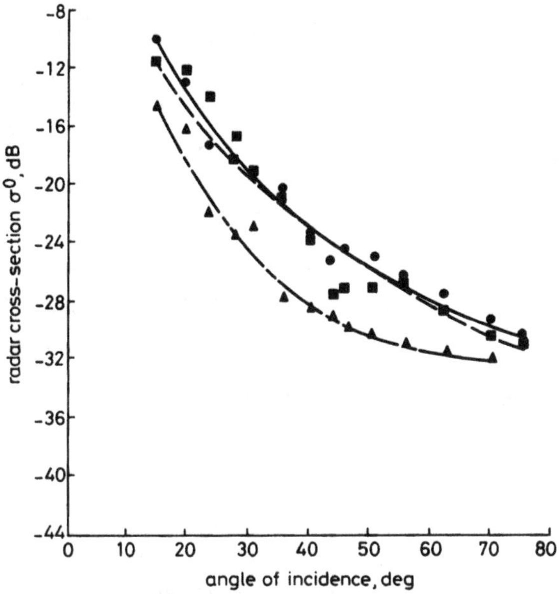

Fig. 2.28 ***Average backscatter coefficient of thick first-year, thin first-year, and lake ice (from Onstott [29], © 1982 IEEE)***
Sensor: TRAMAS
Frequency: 1·5 GHz
Polarisation: VV
●——— thick first-year
■— — thin first-year
▲—— - lake ice

1. At 60° incident angle, the backscatter coefficient of sea ice is about −20 dB. This also corresponds to that of sea clutter in sea state 3 to 4 at 30° grazing angle. At 30° incident angle, the average backscatter coefficient of sea ice is about 20 dB. This corresponds to that of sea clutter in sea state 2 at 60° grazing angle. Unfortunately, there is a lack of data on low grazing angles.

Ogawa *et al.* [30] studied the amplitude distribution of sea ice in the sea of Okhotsk with an X-band radar. The azimuth beamwidth of this radar is 1·2°, and the pulse width is 80 ns.

The data was recorded at midnight on 22 February 1986. Weather conditions were clear and the wind velocity was 3·7 m/s. The direction of wind was south-west and the temperature was −9·3°C.

Data was recorded digitally on the floppy disc of a microcomputer after digitising by an 8 bit A/D convertor. The sampling rate is 25 MHz. The block diagram of the data recording system is shown in Fig. 2.29.

Sea-ice clutter was measured at 143° 22′ east longitude and 44° 21′ north

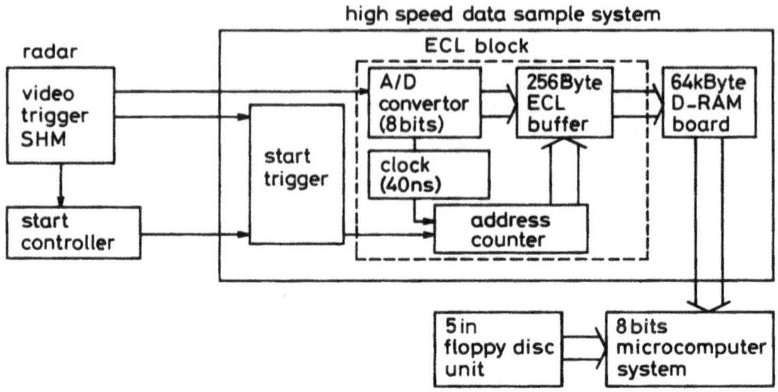

Fig. 2.29 *Block diagram of data reocrding system*

latitude, which is located at the city of Mombetsu in Hokkaido. As shown in Fig. 2.30*a*, the area covered 22·0° to 47·0° in the azimuth direction and 1320 to 2855 m length in the radial direction. The grazing angles were calculated to be 0·53° at 1320 m and 0·24° at 2855 m. Fig. 2.30*b* shows the amplitude of sea-ice clutter against the azimuth and the radial direction. The observed sea ice included fast ice, pack ice and hummock.

To study the amplitude distribution of sea ice, the data was recorded on a 24·6° sector of a single scan. The range interval from 1320 to 2855 m was divided into 256 range bins. Each range bin is equal to 6 m according to the 40 ns sampling period. The azimuth interval between adjacent sweeps is equal to 0·1°, according to the 28 rev/min antenna scan rate and 1680 Hz pulse repetition frequency. Therefore the total data points in this sector are equal to $256 \times 246 = 62\,976$. To obtain the amplitude distributions for different directions, the data was processed in 2·4° subsector. The total data points in this subsector are equal to $256 \times 24 = 6144$. These data were fitted with Weibull distribution. The result is shown in Fig. 2.31 for 34·0° to 36·4°. The straight line in this Figure was drawn by the least-squares method. The deviation of the data points from this straight line was calculated as root-mean-square error (RMSE).

The shape parameter of Weibull distribution can be calculated from the slope of the straight line. The values of shape parameters and RMSE over 22·0° to 46·6° are summarised in Table 2.7.

It can be seen from this Table, that the amplitude distribution of sea-ice clutter can be fitted with Weibull distribution with shape parameter of 0·5 to 1·65.

Although there are overlaps between sectors, the shape parameters vary considerably between adjacent sectors. This phenomena reflects the non-uniform distribution of sea ice in space.

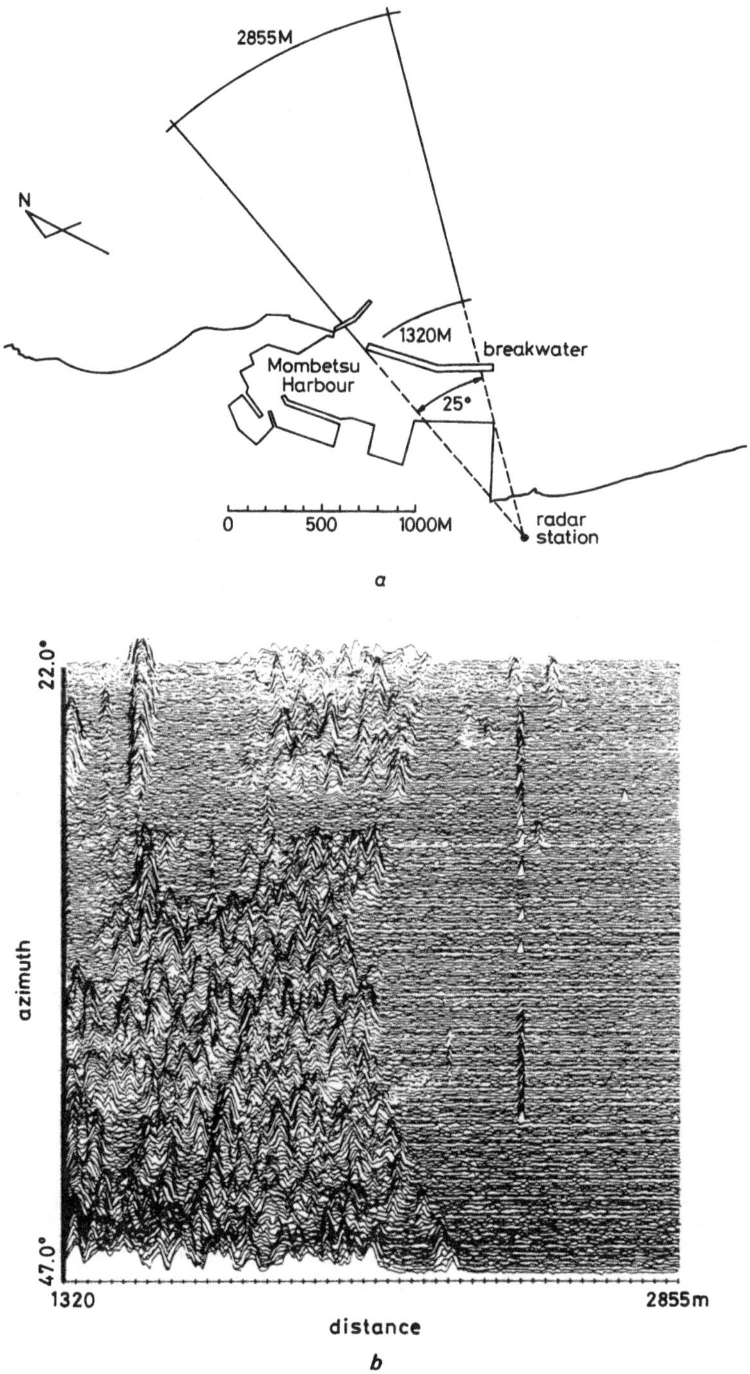

Fig. 2.30 *Observed area of sea ice (a), and the result of observations (b)*

Table 2.7 *Shape parameters for different azimuth sectors*

Sector azimuth (deg)	c	RMSE	Sector azimuth (deg)	c	RMSE
22·0—24·4	0·705	0·2118	33·4—35·8	1·207	0·0821
22·6—25·0	0·697	0·2097	34·0—36·4	1·195	0·0709
23·2—25·6	0·654	0·2480	34·6—37·0	1·211	0·0641
23·8—26·2	0·545	0·2617	35·2—37·6	1·199	0·0740
24·4—26·8	0·501	0·2659	35·8—38·2	1·183	0·1189
25·0—27·4	0·595	0·2463	36·4—38·8	1·145	0·0902
25·6—28·0	0·655	0·1960	37·0—39·4	1·154	0·0804
26·2—28·6	0·697	0·2254	37·6—40·0	1·109	0·0814
26·8—29·2	0·796	0·2438	38·2—40·6	1·109	0·0597
27·4—29·8	1·649	0·2911	38·8—41·2	1·106	0·0592
28·0—30·4	1·085	0·3121	39·4—41·8	1·074	0·0757
28·6—31·0	0·885	0·2235	40·4—42·4	1·037	0·0687
29·2—31·6	0·899	0·1616	40·6—43·0	0·972	0·1017
29·8—32·2	0·927	0·1824	41·2—43·6	0·914	0·1259
30·4—32·8	0·856	0·1959	41·8—44·2	0·874	0·1398
31·0—33·4	0·890	0·1419	42·4—44·8	0·938	0·1651
31·6—34·0	1·010	0·1114	43·0—45·4	1·024	0·1540
32·2—34·6	1·108	0·0956	43·6—46·0	1·040	0·1550
32·8—35·2	1·215	0·0763	44·2—46·6	1·059	0·1281

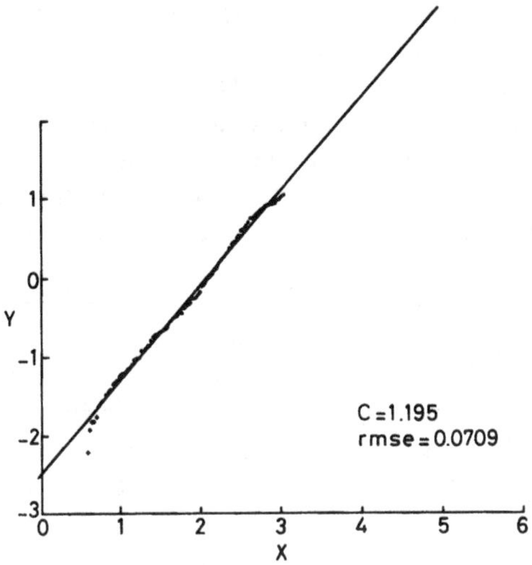

Fig. 2.31 *Data of sea-ice clutter fitted with Weibull distribution*

2.4 Weather clutter

It is well known that weather clutter can be fitted with Rayleigh distribution very well owing to the rather uniformly distributed rain clouds in space. However, when the weather conditions are stormy and windy, the non-homogeneous property of weather clutter become apparent. The amplitude distribution of weather clutter will deviate from Rayleigh distribution and come close to Weibull distribution even for low-resolution radar.

Sekine *et al.* [31—33] reported the Weibull-distributed weather clutter recorded with an L-band air-route surveillance radar (ARSR). The radar parameters are as follows:

Frequency:	1·3 GHz
Antenna horizontal beamwidth:	1·2°
Antenna vertical beamwidth:	3·4°
Elevation angle:	2·9°
Antenna scan rate:	6·0 rev/min
Polarisation:	horizontal
Pulsewidth:	3·0 μs
Pulse repetition frequency:	350 Hz
Transmitted power:	2 MW
Antenna gain:	36·9 dB
Receiver noise factor:	4·0 dB
Total system loss:	7·6 dB

Weather conditions were stormy and windy with a wind speed of 4—14 knots which was measured at the radar site. The wind direction was south-east. Weather clutter was observed from rain clouds in a range interval of 60—65·6 n-miles, over an azimuth interval of 41·8° to 54·2° (range sweep number 400—519). Since the radar beam illuminates the sea surface only over the range interval of 13—27 n-miles, the clutter in the previous range interval contains weather clutter only. Range intervals of 5·6 n-miles between ranges of 60 to 65·6 n-miles were divided into 23 range bins, each corresponding to the pulsewidth.

Data were recorded digitally on magnetic tape, as the inphase component I, and quadrature component Q, video signals, after passing through an IF amplifier and a phase detector. The sample intervals between two adjacent range bins and between two adjacent range sweeps in azmuth direction are 0·25 n-miles and 0·1044°, respectively. Each value of I and Q was recorded on the magnetic tape as a 10 bit signal, and hence the minimum and maximum integer values of I and Q were -512 and $+511$, respectively.

The recorded data for adjacent scans are shown in Figs. 2.32*a* and *b*. From these Figures, storms can be clearly seen and weather clutter has a dynamic range in excess of 20 dB. Over a 10 s period, the amplitude distributions vary greatly with an azimuth interval of range sweep number 400—519.

The recorded data was divided into several sample areas. Each includes 23 range bin (5·6 n-miles) and 20 range sweeps (2°). The number of data points is thus 460. Using these, the parameters of Weibull distribution can be determined. The data from 400—419 range sweeps in five adjacent scans fitted with Weibull distribution are shown in Figs. 2.33*a*—*e*.

The straight line in these Figures can be expressed by

$$Y = cX - c \ln b \tag{2.16}$$

where

$$Y = \ln\left\{-\ln\left[1 - \int_0^x p(x)\,dx\right]\right\} \tag{2.17}$$

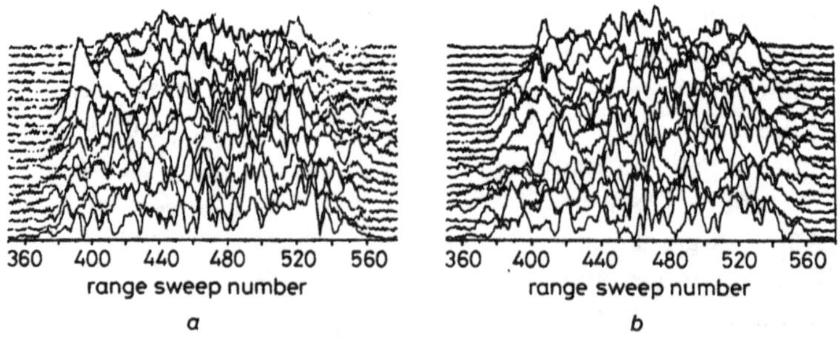

Fig. 2.32 *Amplitude of weather clutter against range sweep number for adjacent scans*

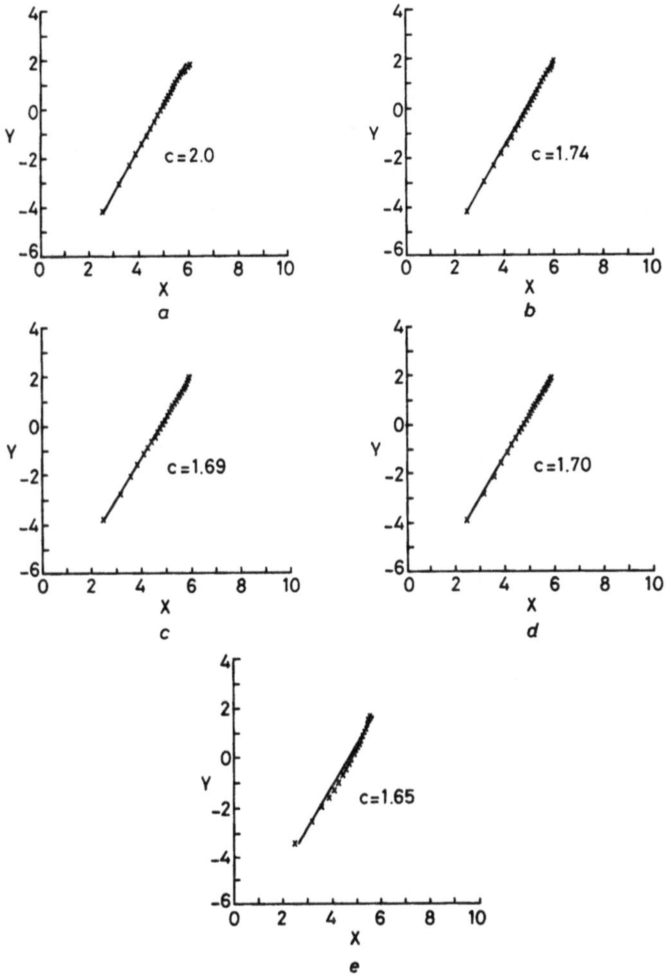

Fig. 2.33 *Weather clutter data fitted with Weibull distribution*

where $p(x)$ is the Weibull probability density function

$$p(x) = \begin{cases} c/b(x/b)^{c-1} \exp[-(x/b)^c] & \text{for } x > 0, b > 0, \text{and } c > 0 \\ 0 & \text{otherwise} \end{cases} \tag{2.18}$$

The slope of this straight line is equal to c. The shape parameters c of Figs. 2.33*a*—*e* are equal to 2·00, 1·74, 1·69, 1·70 and 1·65, respectively.

The values of c for five scans of six sectors of 2° azimuth extent are given in Table 2.8.

It can be seen from this Table that the shape parameter c of weather clutter is very close to 2·00 (Rayleigh distribution). The smallest value of c is equal

Table 2.8 *Shape parameters for different azimuth sectors and scans*

Azimuth (deg.)	Sweep number	Shape parameter c scan 1	scan 2	scan 3	scan 4	scan 5
41·8—43·8	400—419	2·00	1·74	1·69	1·70	1·65
43·9—45·8	420—439	1·92	1·85	1·65	1·86	1·78
45·9—47·9	440—459	2·00	1·97	1·94	1·91	2·00
48·0—50·0	460—479	1·94	1·87	2·00	1·69	1·98
50·1—52·1	480—499	1·91	1·97	1·87	1·85	2·00
52·2—54·2	500—519	2·00	1·98	1·91	1·95	2·00

to 1·65. The average value of c is equal to 1·89. However, the average value of the first sector (sweep number 400—419) is equal to 1·756. This means that, in this region, the storm activity is stronger than in other regions.

2.5 References

1 BALLARD, A. H.: 'Detection of radar signals in log-normal sea-clutter', TRW Sys. Doc. 7425-8509-T0-000, 31 May 1966
2 TRUNK, G. V., and GEORGE, S. F.: 'Detection of targets in non-Gaussian sea clutter', *IEEE Trans.*, 1970, **AES-6**, pp. 620–628
3 BOOTHE, R. R.: 'The Weibull distribution applied to the ground clutter backscatter coefficient'. US Army Missile Command, Technical Report, RE-TR-69-15, AD A691109, 1969
4 JAKEMAN, E., and PUSEY, P. N.: 'A model for non-Rayleigh sea echo'. IEEE Int. Conf. Radar '77, Oct. 1977, pp. 105–109
5 BISHOP, G.: 'Amplitude distribution characteristics of X-band radar sea clutter and small surface targets'. Royal Radar Establishment Memorandum 2348, 1976
6 BARTON, D. K.: 'Target detection in land clutter', Raytheon Company, Wayland, MA, Internal Memorandum, 30 Nov. 1966
7 Working Group: 'Expected performance of SAM–D against SRAM type targets in realistic clutter environments', Raytheon Company, Wayland, MA, Report BR-4356, 21 Aug. 1967
8 NATHANSON, F. E.: 'Sea and land backscatter', Radar Training Program Notes, The Johns Hopkins University Applied Physics Laboratory, chap. 7
9 HOLLIDAY, E. M., WOOD, W. E., POWELL, D. E., and BASHAM, C.E.: 'L-band clutter measurements', US Army Missile Command Report RE-TR-65-1, 3 Nov. 1964
10 LINELL, T.: 'An experimental investigation of the amplitude distribution of radar terrain return', Institute of National Defense, Stockholm, Sweden, Report No. D 3135-62, Oct. 1966
11 CURRIE, N. C., and ZEHNER, S. P.: 'Millimeter wave land clutter model', IEE Int. Conf. Radar '82, Oct. 1982, pp. 385–389
12 BARTON, D. K.: 'Radar clutter' (Artech House, Inc., Dedham, Mass., 1975)
13 MILLER, R.: 'Characterisation of noncoherent ground clutter', Proc. 1984 Int. Symp. on noise and clutter rejection in radars and imaging sensors, ISNCR–84, Oct. 1984, pp. 59–64
14 SEKINE, M., OHTANI, S., MUSHA, T., IRABU, T., KIUCHI, E., HAGISAWA, T., and TOMITA, Y.: 'Weibull distributed ground clutter', *IEEE Trans.*, 1981, **AES-17**, pp. 596–598
15 OLIN, I. D.: 'Amplitude and temporal statistics of sea spike clutter', IEE Int. Conf. Radar '82, Oct. 1982, pp. 198–202
16 EWELL, G. W., TULEY, M. T., and HORNE, W. F.: 'Temporal and spatial behavior of high resolution sea clutter 'spikes' ', IEEE 1984 National Radar Conference, April 1984, pp. 100–104
17 LONG, M. W.: 'Polarization and statistical properties of clutter', Proc. 1984 Int. Symp. on noise and clutter rejection in radars and imaging sensors, Oct. 1984, pp. 25–32
18 Johns Hopkins University, Silver Spring, Md., 'Sea clutter model, SCM-1', 1970 Nov. 25 (Unpublished Report)
19 SCHLEHER, D. C.: 'Radar detection in Weibull clutter', *IEEE Trans.*, 1976, **AES-12**, pp. 736–743
20 FAY, F. A., CLARKE, J., and PETERS, R. S.: 'Weibull distribution applied to sea clutter', IEE Int. Conf. Radar '77, pp. 101–104
21 SEKINE, M., MUSHA, T., TOMITA, Y., HAGISAWA, T., IRABU, T., and KIUCHI, E.: 'Weibull distributed sea clutter', *IEE Proc.*, 1983, **130F**, p. 476
22 HAVIG, T., and CHIN, P. W.: 'Private communication to F. A. Fay, J. Clarke and R. S. Peters, 1977
23 MAALOE, J.: 'Sea clutter statistics', IEE Int. Conf. Radar '82, Oct. 1982, pp. 193–197

24 SEKINE, M., MUSHA, T., TOMITA, Y., HAGISAWA, T., IRABU, T., and KIUCHI, E.: 'Log-Weibull distributed sea clutter', *IEE Proc.*, 1980, **127F**, pp. 225–228

25 TRIZNA, D. B.: 'Open ocean radar sea scatter measurements', IEEE 1985 Int. Radar Conf., May 1985, pp. 135–140

26 KALMYKOV, A. I., and PUSTOVOYTENKO, V. V.: 'On polarization features of radio signal scattered from the sea surface at small grazing angles', *J. Geophysical Research*, 1976, **8**, pp. 1960–1968.

27 KALMYKOV, A. I., KUREKIN, A. S., LAMENTA, Yu. A., OSSROVSKII, I. E., and PUSTOVOYENKO, V. V.: 'Characteristics of microwave scattering from breaking sea waves', *Translation of Radiophysics*, 1976, **19**, pp. 1315–1321

28 LEWIS, B. L., and OLIN, I. D.: 'Experimental study and theoretical model of high resolution radar backscatter from the sea', *Radio Science*, 1980, **15**, pp. 815–828

29 ONSTOTT, R. G., MOORE, R. K., GOGINENI, S., and DELKER, C.: 'Four years of low-altitude sea ice broad-band backscatter measurements', *IEEE J.*, 1982, **OE-7**, pp. 44–50

30 OGAWA, H., SEKINE, M., MUSHA, T., AOTA, M., OHI, M., and FUKUSHI, H.: 'Weibull-distributed radar clutter reflected from sea ice', *Trans. IEICE*, 1987, **E70**, pp. 116–120

31 SEKINE, M., MUSHA, T., TOMITA, Y., HAGISAWA, T., IRABU, T., and KIUCHI, E.: 'On Weibull-distributed weather clutter', *IEEE Trans.*, 1979, **AES-15**, pp. 824–830

32 SEKINE, M., MUSHA, T., TOMITA, Y., HAGISAWA, T., IRABU, T., and KIUCHI, E.: 'Suppression of Weibull-distributed weather clutter', IEEE International Radar Conference, April 1980, pp. 294–298

33 SEKINE, M., MUSHA, T., IRABU, T., KIUCHI, E., HAGISAWA, T., and TOMITA, Y.: 'Non-Rayleigh weather clutter', *IEE Proc.*, 1980, **127F**, pp. 471–474

34 LONG, M. W.: 'Radar reflectivity of land and sea' (Artech House, Inc., Dedham, Mass., 1983)

Chapter 3

Signal detection in Weibull clutter

The theory of radar signal detection was founded by Rice, Marcum, Swerling and others since the Second World War. However, most of these theories are based on signal detection in Gaussian non-correlated noise. This is valid only in the receiver-noise case. When the radar signal is embedded in clutter, it is a situation of signal detection in non-Gaussian correlated noise. Until the 1970s, Trunk [1, 2, 3] discussed signal detection in non-Gaussian and log-normal clutter. Signal detection in Weibull clutter was first carried by Goldstein [4], Ekstrom [5] and Schleher [6], and recently by Farina *et al.* [7].

There are two kinds of signal detection: signal detection in time domain, and signal detection in frequency domain. In the early stage of development of radar technology, signal detection in time domain was widely used, the simplest being threshold crossing detection. Binary moving-window detectors are still widely used for non-coherent integration. However, signal detection in the time domain is effective only in a receiver-noise environment, but not in a clutter environment. It only has super-clutter visibility, i.e. signal detection in a signal-to-clutter ratio greater than 0 dB.

Signal detection in the frequency domain not only uses the amplitude difference between signal and clutter, but also uses the difference in Doppler frequency between signal and clutter. Therefore, it can obtain sub-clutter visibility while associated with MTI or MTD techniques to reject clutter before detection. Since the MTI filter is a comb filter, however, it can output a moving-target signal still in time domain after cancelling the clutter. Therefore, strictly, only the MTD system is a signal-detection system in the frequency domain.

Knowledge of amplitude distribution of clutter is important not only for signal detection in the time domain, but also for signal detection in the frequency domain. This is due to the fact that the transform from time domain to frequency domain, such as FFT, is a linear transform; so the amplitude distribution will not be changed after the transform. The problem of how to maintain the false-alarm rate constant is important, whether in signal detection in time domain or in signal detection in frequency domain.

The design of a CFAR detector, whether in the time domain or frequency domain, depends on a knowledge of the amplitude distribution of the clutter.

In this Chapter, we will discuss the false-alarm probability versus threshold for Weibull clutter. Then we will be concerned with the design of CFAR detector for Weibull-distributed clutter. Finally, we will discuss the non-parametric, i.e., distribution-free, CFAR detector and its performance.

3.1 False-alarm probability in Weibull clutter

How to control the false-alarm rate of a detector is an important problem in radar signal detection. If the false-alarm rate is too high, it will block the data-processing computer. Otherwise, if the false-alarm rate is too low, it will degrade the sensitivity of signal detection. Therefore, it is often desired to design a detector with constant false-alarm rate (CFAR) capability.

The false-alarm rate of a detector depends on the distribution function of the clutter on one hand, and on the detection algorithm on the other. In general, before designing a detector with good CFAR performance, we must have sufficient knowledge about the amplitude distribution of the clutter.

It is well known that the relationship between false-alarm probability and threshold to mean for Weibull clutter can be calculated from

$$P_{fa} = \int_T^{\infty} p(v)\,dv = \exp\left\{-\left[\frac{T}{b}\right]^c\right\} \tag{3.1}$$

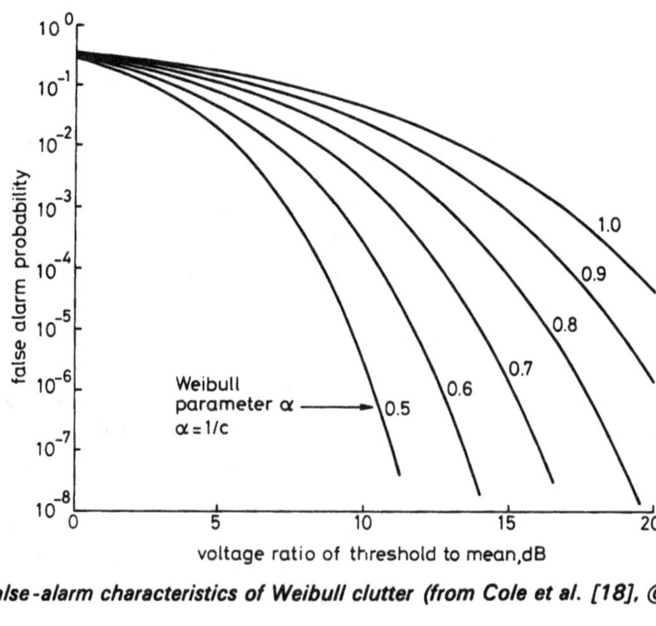

Fig. 3.1 *False-alarm characteristics of Weibull clutter (from Cole et al. [18], © 1978 IEEE)*

where T is the voltage ratio of threshold to mean, c is the shape parameter, and b is the characteristic value.

This result is plotted in Fig. 3.1 [18].

It can be seen from Fig. 3.1 that the false-alarm probability not only depends on the voltage ratio of threshold to mean, but also on the shape parameter c. As previously mentioned, the radar clutter will have different shape parameters for different space location. If the threshold-to-mean ratio is set originally to be 10·4 dB for a false-alarm probability of 10^{-6} and for Weibull clutter with $c = 2{\cdot}0$ (Rayleigh distribution), a slight change in shape parameter from $c = 2{\cdot}0$ to 1·67 will yield a hundred-fold increase in the false-alarm probability. This increase in the false-alarm rate is likely to overload a radar detection system. Therefore, how to design a CFAR detector in a Weibull clutter environment is an important problem in modern radar design.

3.2 CFAR detector for Weibull clutter

It is well known that the conventional cell-averaging CFAR detector estimates the mean value with a limited number of reference cells. This type of CFAR detector can maintain the false-alarm rate constant only for Rayleigh-distributed clutter and with sufficient number of reference cells. Since for Rayleigh-distribution only one parameter has to be estimated, once the mean has been estimated, the whole function can be evaluated. However, since Weibull distribution is a two-parameter (mean and shape parameter) distribution function, the cell-averaging process can estimate the mean only, but cannot estimate the shape parameter of the distribution. Hansen [8] showed the performance of a cell-averaging CFAR detector in Weibull clutter (see Fig. 3.2). The dashed line is an ideal threshold characteristic for Rayleigh distribution with an exponential function. The full line nearby is the threshold characteristic of a cell-averaging CFAR detector for Rayleigh clutter. The other three curves are threshold characteristics for Weibull distribution with different shape parameters. If the shape parameter cannot be known *a priori*, the cell-averaging CFAR detector cannot maintain the false-alarm rate constant.

Hansen [9] pointed out that a generalised CFAR detector can be constructed with an off-line estimator and an on-line zero-memory non-linear filter. The off-line estimator estimates the unknown distribution parameters (including the scale and shape parameters) from an appropriate set of observations. The on-line zero-memory non-linear filter is controlled by these estimates so that an output is obtained which, for any input noise belonging to the given class, yields an output with a known and normalised probability density function (PDF).

Assume it has been decided that the class of PDFs $p(x; \nu, \eta, \ldots, \gamma)$ will

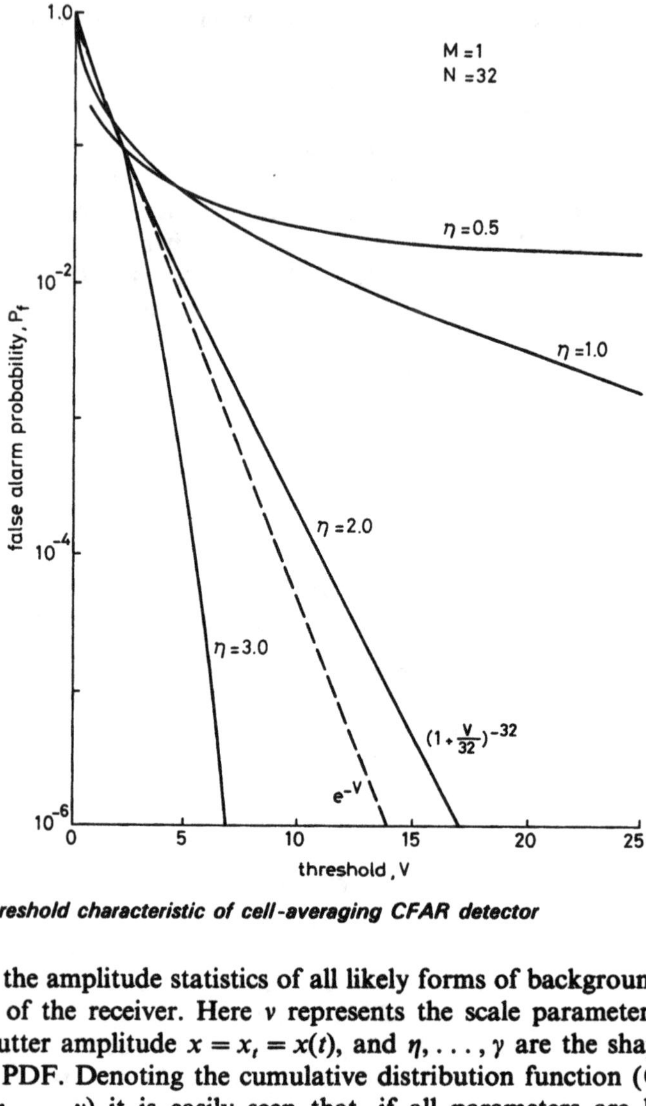

Fig. 3.2 *Threshold characteristic of cell-averaging CFAR detector*

encompass the amplitude statistics of all likely forms of background clutter at the output of the receiver. Here v represents the scale parameter associated with the clutter amplitude $x = x_t = x(t)$, and $\eta, \ldots, \gamma$ are the shape parameters of the PDF. Denoting the cumulative distribution function (CDF) of x, as $F(x; v, \eta, \ldots, \gamma)$ it is easily seen that, if all parameters are known, the zero-memory, non-linear transformation

$$z(t) = -\ln[1 - F(x; v, \eta, \ldots, \gamma)] \tag{3.2}$$

will transfrom the clutter $x(t)$ into an output $z(t)$ which has a normalised exponential PDF

$$p(z) = \begin{cases} \exp(-z) & z \geqslant 0 \\ 0 & \text{otherwise} \end{cases} \tag{3.3}$$

The basic problem with this approach is that the actual values of the parameters $v, \eta, \ldots, \gamma$ are not known in advance. We are thus led to the strategy of attempting to estimate their values on the basis of an appropriate set of reference observations. The estimation procedure to be used must satisfy several requirements such as quality of the estimates, CFAR performance, and the complexity of the resulting implementation. No general procedure has been determined for deriving a 'best' estimation procedure. In the following we illustrate with an example the design of a generalised CFAR detector for Weibull clutter and evaluate its performance.

The Weibull class of probability desity function has the form

$$p(x; v, \eta) = \begin{cases} \dfrac{\eta}{v}\left(\dfrac{x}{v}\right)^{\eta-1} \exp\left[-\left(\dfrac{x}{v}\right)^{\eta}\right] & x \geqslant 0 \\ 0 & \text{otherwise} \end{cases} \tag{3.4}$$

where v is a scale parameter and η is a shape parameter. Thus if we have a linear receiver followed by an envelope detector the case of Gaussian noise corresponds to $\eta = 2$ and $v^2 = 2\sigma^2$. The CDF corresponding to (3.4) is

$$F(x; v, \eta) = \begin{cases} 1 - \exp[-(x/v)^{\eta}] & x \geqslant 0 \\ 0 & \text{otherwise} \end{cases} \tag{3.5}$$

Inserting this CDF into (3.2) then leads to the following expression for the required zero-memory non-linear transformation:

$$\begin{aligned} z(t) &= -\ln\{\exp[-(x(t)/v)^{\eta}]\} \\ &= (x(t)/v)^{\eta} \end{aligned} \tag{3.6}$$

To avoid the η-th law device in a practical system we may write

$$z(t) = \exp\{\eta[\ln(x(t)) - \ln v]\} \tag{3.7}$$

which can be implemented as shown in Fig. 3.3.

In order to estimate the parameters v and η we shall assume that a set of N reference-noise observations are taken in the time domain around the position presently under test for the presence of a target. The set of reference-noise observations is denoted $x_1, x_2, \ldots, x_N$. Several procedures can be used for estimating the unknown parameters v and η. The maximum likelihood procedure is very cumbersome to use whereas the method of moments leads to

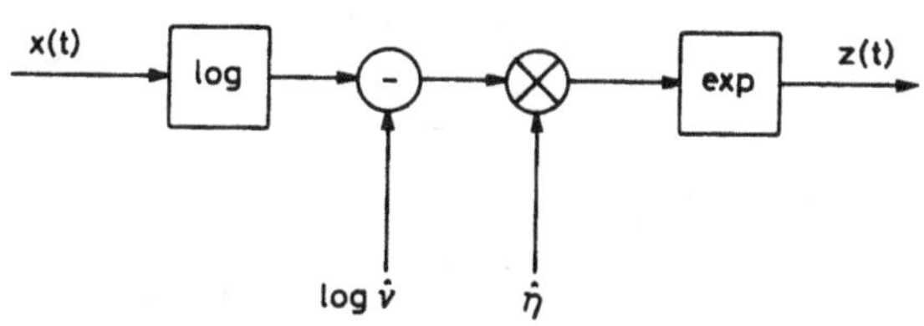

Fig. 3.3 *The generalised CFAR detector for Weibull clutter*

simple expressions for the estimates [10]. An estimation procedure proposed in Ref. 11 is particularly attractive because it leads to a perfect constant false-alarm rate for all values of the parameters ν and η. From Ref. 11 the estimate of η is

$$\hat{\eta} = \left\{\frac{6}{\pi^2}\frac{N}{N-1}\left[\frac{1}{N}\sum_{i=1}^{N}(\ln x_i)^2 - \left(\frac{1}{N}\sum_{i=1}^{N}\ln x_i\right)^2\right]\right\}^{-1/2} \tag{3.8}$$

and the estimate of ν is

$$\nu = \exp\left\{\frac{1}{N}\sum_{i=1}^{N}\ln x + 0{\cdot}5772\hat{\eta}^{-1}\right\} \tag{3.9}$$

It is not difficult to show that the use of (3.8) and (3.9) in the off-line estimator results in an output $z(t)$ which has a distribution which is independent of the true value of ν and η.

Hansen [8] suggested a practical CFAR detector for Weibull clutter based on the method of moments. The resulting design of this Weibull CFAR detector is shown in Fig. 3.4. It uses the following test statistic:

$$t = \exp\left\{\frac{\lg V_i - \frac{1}{N}\sum_{i=1}^{N}\lg V_i}{\left[\frac{1}{N}\sum_{i=1}^{N}\lg V_i\right]^2 - \frac{1}{N}\sum_{i=1}^{N}(\lg V_i)^2}\right\} \overset{?}{>} T \tag{3.10}$$

The performance of this CFAR detector in Weibull clutter has been simulated. The results are shown in Fig. 3.5. For each value of N the curves, which were obtained by computer simulation using the important sampling technique, were identical for all values of the Weibull parameters. The CFAR loss of this detector was determined for the case of a stationary Gaussian noise input

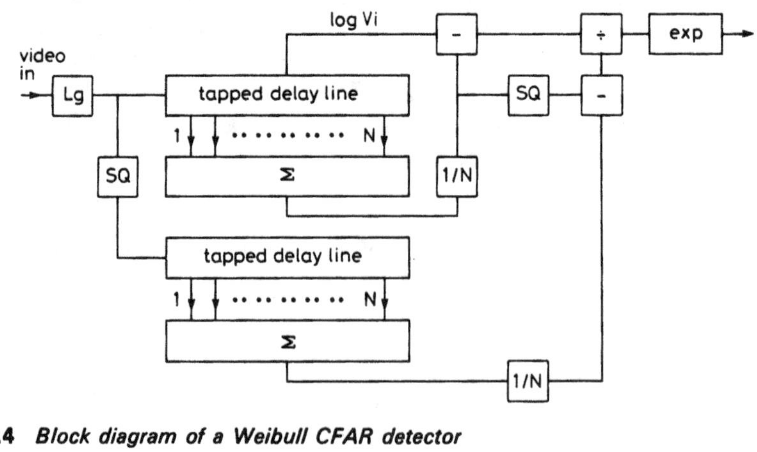

Fig. 3.4 *Block diagram of a Weibull CFAR detector*

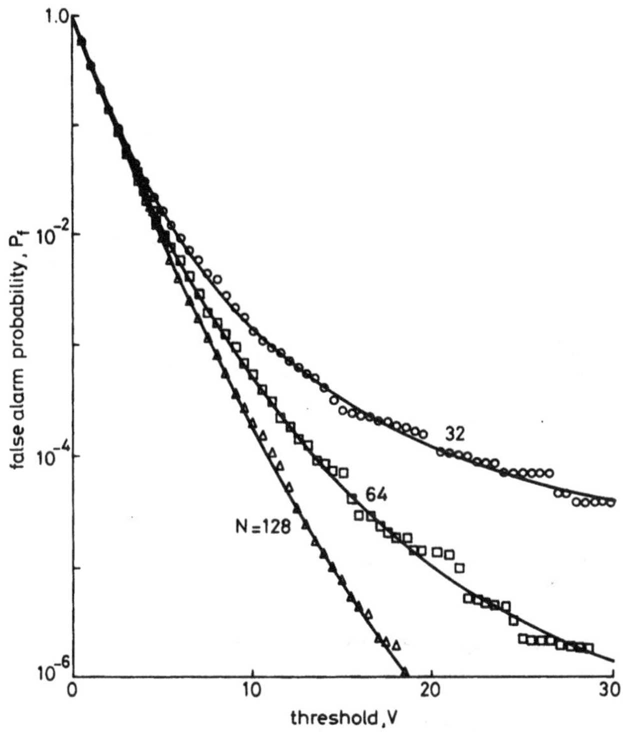

Fig. 3.5 *Threshold characteristic*

which, at the output of an envelope detector, leads to a Rayleigh PDF. These results are shown in Fig. 3.6.

A CFAR detector which maintains a constant false-alarm rate in Weibull clutter has been proposed by Goldstein [4]. In this CFAR detector a test statistic termed log t test is constructed.

$$t = \frac{\ln V_o - \frac{1}{N}\sum_{i=1}^{N} \ln V_i}{\sqrt{\frac{1}{N}\sum_{j=1}^{N}\left[\ln V_j - \frac{1}{N}\sum_{k=1}^{N} \ln V_k\right]^2}} \overset{?}{>} T \tag{3.11}$$

where V is the envelope detected voltage and N is the number of reference cells. Notice that, when only clutter is present, the numerator of t involves the subtraction of the maximum-likelihood estimate of the mean of $\ln V_o$ and the denominator of t normalises the variable by dividing by the maximum-likelihood estimate of the standard deviation of $\ln V_o$. The test for the presence of signal consists in determining whether or not t in (3.11) is greater than a fixed level T.

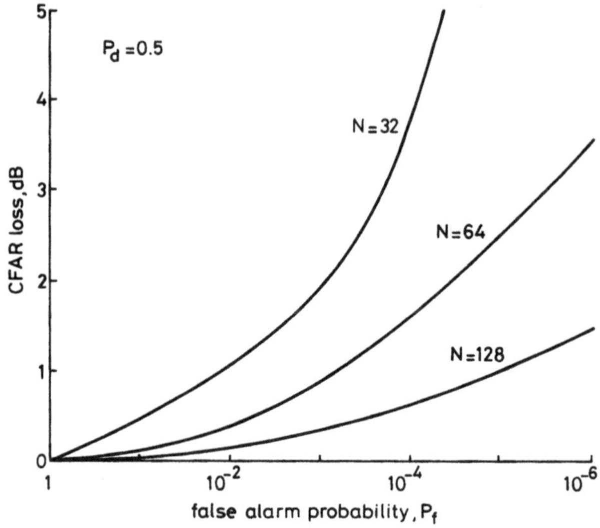

Fig. 3.6 *CFAR loss of the Weibull CFAR detector*

The false-alarm probability P_{fa} and detection probability P_d are given by

$$P_{fa} = \text{prob.}\{t > T | \text{ clutter only}\} = \int_T^\infty p_c(t)\,dt \tag{3.12}$$

and

$$P_d = \text{prob.}\{t > T | \text{ signal + clutter}\} = \int_T^\infty p_{s+c}(t)\,dt \tag{3.13}$$

where $p_c(t)$ denotes the probability function of the test statistics when only clutter is present, and $p_{s+c}(t)$ denotes the probability density function when signal plus clutter is present.

In principle, the distribution of the log t statistic when the target is absent and the clutter is Weibull can be calculated. However, the resulting expressions have been found to be unwieldly (requiring the numerical evaluation of multiple integrals), except for the special case in which only two cells are employed ($N = 2$). Normally, the loss in detection efficiency associated with attempting to design a CFAR detector using only two reference cells would exclude this case from practical interest. Therefore, the following approach is used.

For N sufficiently large, the quantities $\hat{m}$ and $\hat{\sigma}$ defined by

$$\hat{m} \triangleq \frac{1}{N}\sum_{i=1}^{N} \ln V_i \tag{3.14}$$

$$\hat{\sigma} \triangleq \frac{1}{N}\sum_{i=1}^{N} (\ln V_i - \hat{m})^2 \tag{3.15}$$

are sufficiently good estimates of their respective true values to justify approximating the residuals by zero mean Gaussian random variables. This permits us to obtain an approximate distribution of the log t statistic in Weibull clutter which is valid when $N \gg 1$.

From Ref. 12, p. 237, the distribution of $\hat{m}$ is asymptotically Gaussian with a mean value of $E\{\ln V\}$ and a variance equal to $\text{var}(\ln V)/N$. It also can be shown [12] that the asymptotic distribution of $\hat{\sigma}^2$ is chi-square in N degrees of freedom. When N is large, the chi-square distribution may be approximated by a Gaussian distribution having the same first two moments. This means that, for a large enough N, we may approximate $\hat{m}$ and $\hat{\sigma}$ as Gaussian random variables. We shall assume a jointly Gaussian distribution for $\hat{m}$ and $\hat{\sigma}$ and utilise the actual correlation between $\hat{m}$ and $\hat{\sigma}$.

In Ref. 12, p. 348 the correlation between $\hat{m}$ and $\hat{\sigma}$ is calculated, allowing the sampled random variables to have any arbitrary probability distribution. The result expressed in terms of the observables at hand (for large N) is

$$\rho(\hat{m}, \hat{\sigma}) = \frac{E\{\ln V - E(\ln V)\}^3}{2\sigma_o N} = -\frac{0{\cdot}234}{N} \tag{3.16}$$

The other relevant moments required are (to order $1/N$)

$$E(\hat{m}) = E(\ln V) = 0{\cdot}058 \triangleq m_o$$

$$E(\hat{\sigma}) = \text{var}(\ln V) = \frac{\pi}{24} \triangleq \sigma_o$$

$$\text{var}(\hat{m}) = \frac{\pi^2}{(24N)}$$

$$\text{var}(\hat{\sigma}) = \frac{1}{32N\sigma_o^2}\left\{[\psi'(1)]^2 + \left[\frac{\psi'''(1)}{2}\right]\right\} = 0{\cdot}45/N \tag{3.17}$$

where $\psi^{(n)}(1)$ denotes the nth derivative of the Euler ψ function. Specific values are $\psi'(1) = \pi^2/6$, $\psi'''(1) = \pi^4/15$.

If we now express the decision statistic in terms of $\hat{m}$ and $\hat{\sigma}$, we see that the false-alarm probability is based upon the test $V_o \geqslant \exp\{\hat{m} + \hat{\sigma}T\}$. Therefore, the false-alarm probability is

$$P_{fa} = \int_{\hat{m}}\int_{\hat{\sigma}} p(\hat{m}, \hat{\sigma}) \int_{\exp[\hat{m} + \hat{\sigma}T]} p(V_o)\, dV_o\, d\hat{m}\, d\hat{\sigma} \tag{3.18}$$

The integration over V_o may be performed in closed form, using the bivariate Gaussian distribution for $\hat{m}$ and $\hat{\sigma}$, Eqn. 3.18 may be simplified to

$$P_{fa} = \frac{1}{\sqrt{2\pi}\sigma_z} \int_{z=-\infty}^{\infty} \exp\left\{-\frac{1}{2}\left[e^z + \frac{(z-\bar{z})^2}{\sigma_z^2}\right]\right\} dz \tag{3.19}$$

where

$$\bar{z}_2 = 2(m_o + \sigma_o T)$$

$$\sigma_o = 4\{\mathrm{var}(\hat{m}) + T^2\,\mathrm{var}(\hat{\sigma}) + 2T\rho(\hat{m}, \hat{\sigma})\} \tag{3.20}$$

When (3.17) is used in conjunction with (3.19), the threshold T may be found as a function of N in order to achieve a desired P_{fa}. To do this, of course, requires the evaluation of the integral in (3.19). Use of the method of saddle-point integration (which becomes exact as $N \to \infty$) yields

$$P_{fa} = \frac{\exp\{-\frac{1}{2}[1 + \frac{1}{4}\sigma_z^2 e^{z_o}]e^{z_o}\}}{\sqrt{1 + \frac{1}{2}\sigma_z^2 e^{z_o}}}, \quad N \gg 1 \tag{3.21}$$

where z_o is the solution of the transcendental equation

$$z_o + \tfrac{1}{2}\sigma_z^2 e^{z_o} = 2(m_o + \sigma_o T) \tag{3.22}$$

Note that, as $N \to \infty$, $\sigma_z^2 \to 0$ and $z_o \to 2(m_o + \sigma_o T_\infty)$, where T_∞ is the asymptotic threshold, which is also the threshold for optimum envelope detector.

A typical Weibull threshold characteristic for the log t test is shown in Fig. 3.7.

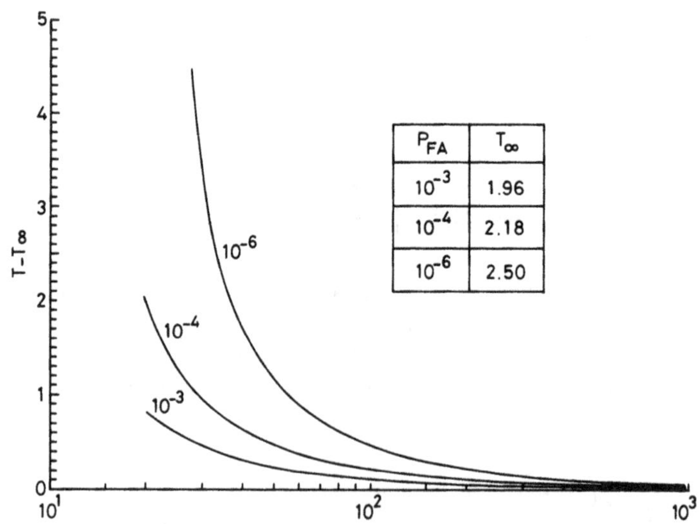

Fig. 3.7 *Threshold characteristic of log t test in Weibull clutter (from Goldstein [4], © 1973 IEEE)*

Simulations have been conducted on a limited basis which indicate that approximate development of the threshold characteristic for $N \gg 20$ and $P_{fa} \geqslant 10^{-4}$. For smaller P_{fa}, N must be larger before the asymptotic results can be relied on.

It can be shown that the log t test has the CFAR property in Weibull clutter. If we write the decision statistic directly in terms of the matched filter output, we have

$$t = \frac{\ln\left[\prod_{j=1}^{N} (V_o/V_j)\right]}{\sqrt{\frac{1}{N}\sum_{k=1}^{N}\left\{\ln\left[\prod_{i=1}^{N}\left(\frac{V_k}{V_i}\right)\right]\right\}}} \tag{3.23}$$

From the form of (3.23) it is seen by inspection that, if $\{V_i; i=0, \ldots, N\}$ were replaced by $\{\alpha V_i^{\beta}; i=0, \ldots, N\}$ (i.e., every V_i is replaced by a version which is raised to the power β and also multiplied by the scale factor α), the quantity t would remain totally unaffected. The parameters α and β permit the class of Weibull random variables to be generated from a class of Rayleigh random variables. The fact that t is functionally independent of α and β establishes the CFAR property for Weibull clutter.

When the clutter is log-normal distributed, it is very easy to show the CFAR property of the log t test. If we let the envelope-detected log-normal clutter pass through an ideal log amplifier, the resulting distribution is Gaussian. From Ref. 12, p. 382 it follows that the log t test statistic in (3.11) has a Student-t probability distribution when the clutter is log-normal. The false-alarm probability in this case is

$$P_{fa} = \int_{T'}^{\infty} f_s^{(N-1)}(v)\, dv \tag{3.24}$$

where

$$T' = \sqrt{\frac{N-1}{N+1}}\, T$$

and where $f_s^N(\cdot)$ denotes the Student-t probability density function in N degrees of freedom.

Note that, the false-alarm probability depends upon T and the number of reference cells N. Since T is independent of clutter parameter, the log t test has the CFAR property in log-normal clutter. Typical threshold characteristics are shown in Fig. 3.8 for the log-normal clutter environment.

If we compare the value of T_∞ for log-normal clutter with the value of T_∞ for Weibull clutter, for the same P_{fa}, it is found that, for any $P_{fa} < 2{\cdot}2 \times 10^{-1}$ (i.e. practically all cases of interest), the asymptotic threshold required in log-normal clutter will be higher than that required for Weibull clutter.

However, as N decreases from $N=\infty$, the threshold characteristic for

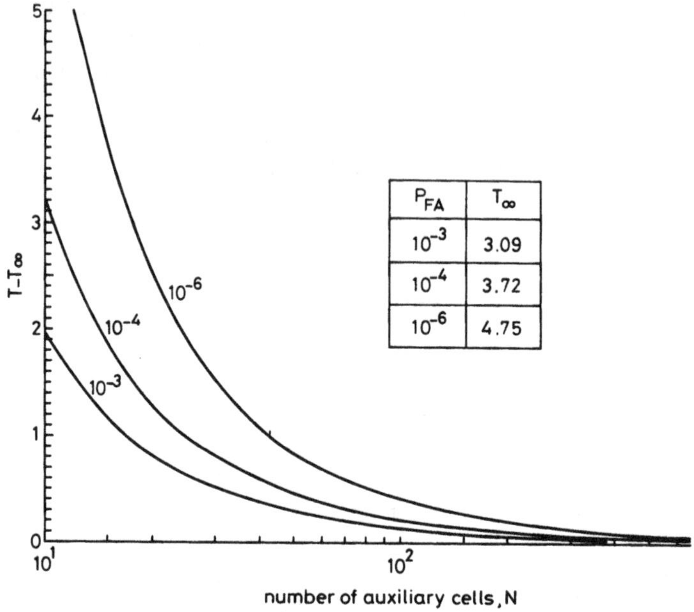

Fig. 3.8 *Threshold characteristics of log t test in log-normal clutter (from Goldstein [4], © 1973 IEEE)*

Weibull clutter departs more rapidly from its asymptotic value than it does in log-normal clutter. Therefore, without actually evaluating the Weibull threshold characteristic for the difficult intermediate value of N, we may state that, for any specified $P_{fa} < 2{\cdot}2 \times 10^{-1}$ the threshold characteristics for log-normal clutter and Weibull clutter will intersect at some value of $N \triangleq N_o$. The significance of N_o is that, if the number of reference cells employed is N_o, then only one value of T need be chosen and the same P_{fa} can be maintained over both the log-normal and Weibull family of clutter distributions. The required value of N increases with decreasing P_{fa}. Without calculating the Weibull threshold characteristic for the intermediate values of N, it is not possible to determine N_o. However, the available curves (Fig. 3.7 and 3.8) indicate that, for example, when $P_{fa} = 10^{-6}, N_o \cong 30$, while for $P_{fa} = 10^{-4}, N_o \cong 18$. For any given P_{fa} or N, the required threshold for log-normal distribution differs drastically from that in Weibull clutter. The distribution of real clutter may be changed from Weibull to log-normal distribution. Therefore, the best way is to use a pattern-recognition network. The purpose of pattern recognition is to determine to which clutter model a given clutter sample belongs.

Szajnowski [13] proposed a method for discrimination between log-normal and Weibull clutter. The principle of this method was suggested by Dumonceaux *et al.* [14].

The log-normal distribution is a function of two independent parameters and its probability density function (PDF) may be written as

$$p_o(x; m, \sigma) = (1/x\sigma\sqrt{2\pi}) \exp[-(\ln x - m)^2/2\sigma^2], \quad x > 0, \sigma > 0 \quad (3.25)$$

where m and σ are the mean value and standard deviation of ln x, respectively.

The Weibull family of distributions, like the log-normal, is a two-parameter family with PDF of the form

$$p_1(x; c, b) = (c/b)(x/b)^{c-1} \exp[-(x/b)^c], \quad x > 0, b > 0, c > 0 \quad (3.26)$$

where b is the scale parameter and c is the shape parameter. The Rayleigh and the exponential distributions are special cases of (3.26) for $c = 2$ and $c = 1$, respectively.

If we let $Z = \ln X$, when X is distributed according to (3.25), the resulting PDF is Gaussian, i.e.

$$p_o(z; m, \sigma) = (1/\sigma\sqrt{2\pi}) \exp[-(z - m)^2/2\sigma^2], \quad \sigma > 0 \quad (3.27)$$

If X has the Weibull distribution (3.26), then $Z = \ln X$ has a type 1 extreme-value distribution with PDF

$$p_1(z; a, d) = (1/d) \exp[(z - a)/d - \exp[(z - a)/d], \quad d > 0 \quad (3.28)$$

The scale and shape parameters, b and c, of Weibull distribution are related to a and d of the type 1 extreme-value distribution by

$$b = \exp(a) \quad (3.29)$$

$$c = 1/d \quad (3.30)$$

Let $Z = (Z_1, Z_2, \ldots, Z_N)$ be a random vector formed from N independent and identically distributed clutter samples. Now suppose that we have observed the realisation z of the random vector Z, and on the basis of this observation we are to decide which of the following hypotheses is true:

H_0: z belongs to the normal distribution

H_1: z belongs to the type 1 extreme-value distribution

We assume that the parameters of both of the distributions are unknown. Thus the hypotheses H_0 and H_1 are composite hypotheses.

When hypothesis H_0 is true, the conditional PDF for the random vector Z is given by

$$p_0(z; m, \sigma) = (2\pi\sigma^2)^{-N/2} \prod_{i=1}^{N} \exp[-(z_i - m)^2/2\sigma^2] \quad (3.31)$$

and when hypothesis H_1 is true, the conditional PDF is of the form

$$p_1(z; a, d) = d^{-N} \prod_{i=1}^{N} \exp\{(z_i - a)/d - \exp[(z_i - a)/d]\} \quad (3.32)$$

Thus we must determine whether the sample vector z belongs to the distribution $p_0(z; m, \sigma)$ or to the distribution $p_1(z; a, d)$.

The ratio of maximised likelihoods (RML) provides a good test for discrimination between two distributions with unknown locations and scale parameters. Using the conditional PDFs, (3.31) and (3.32), the RML can be written as

$$\Lambda(z) = \frac{\max_{a,d} \left(d^{-N} \prod_{i=1}^{N} \exp\{(z_i - a)/d - \exp[(z_i - a)/d]\} \right)}{\max_{m,\sigma} \left\{ (2\pi\sigma^2)^{-N/2} \sum_{i=1}^{N} \exp[-(z_i - m)^2/2\sigma^2] \right\}} \tag{3.33}$$

where $z_i, i = 1, 2, \ldots, N$, is the ith component of a sample vector z. This method directs us toward making maximum-likelihood estimates (MLE) of the unknown parameters and to substituting these estimates in the likelihood functions. The estimates are therefore used in the same manner as the true parameter values which had been known.

Unfortunately, MLEs of type 1 extreme-value distribution parameters a and d must be determined by iterative methods which are not well suited to hardware implementation. Szajnowski [13] suggested using estimates obtained by the method of moments. Hence we may use the following estimates:

$$a^* = \hat{m} + \gamma d^* \tag{3.34}$$

$$d^* = \hat{\sigma}\sqrt{6}/\pi \tag{3.35}$$

where γ is the Euler constant, and the estimates $\hat{m}$ and $\hat{\sigma}^2$ are given by

$$\hat{m} = (1/N) \sum_{i=1}^{N} z_i \tag{3.36}$$

$$\hat{\sigma}^2 = (1/N) \sum_{i=1}^{N} (z_i - \hat{m})^2 \tag{3.37}$$

Substituting a^* and d^* into (3.33) we get the expression for the ratio of 'almost' maximised likelihoods, which, on taking logarithms and after some algebraic manipulations, gives

$$\begin{aligned} \ln L(z) &= (N/2)\ln(\pi^3/3) - N\gamma + N/2 \\ &\quad - \exp(-\gamma) \sum_{i=1}^{N} \exp[\pi(z_i - \hat{m})/\hat{\sigma}\sqrt{6}] \end{aligned} \tag{3.38}$$

We can therefore consider

$$r(z) = -(1/N) \sum_{i=1}^{N} \exp\{\pi(z_i - \hat{m})/\hat{\sigma}\sqrt{6}] \tag{3.39}$$

to be the r-test statistic for deciding whether the clutter distribution is log-normal or Weibull. The decision rule may be stated as follows: choose

$p_1(z; a, d)$ as a model if

$$r(z) > r_c \tag{3.40}$$

and choose $p_0(z; m, \sigma)$ otherwise; r_c is the critical value of the r-test.

Further, an additional approximation may now be made in order to obtain a procedure more suitable for hardware implementation. Using the power expansion for the exponential function in eqn. 3.39 and taking the first four terms, we may define

$$d(z) = -(1/N) \sum_{i=1}^{N} [(z_i - \hat{m})/\hat{\sigma}]^3 \tag{3.41}$$

to be the d-test statistic. We reject or accept H_0, depending on whether $d(z)$ is greater than or less than a predetermined critical value d_c.

The problem is how to determine the critical values of r and d_c. These are determined by the permissible error of discrimination. There are two types of error associated with any decision rule in a two-class problem. If H_0 is in fact true, and the decision is to reject H_0, then an error, called type I error, has been committed. On the other hand, if H_1 is in fact true, and the decision is to accept H_0, then an error, called the type II error, has been committed. The probability of committing a type I error is denoted by α, while the probability of committing the type II error is denoted by β. Usually α is fixed at some small value and the 'goodness' of the decision rule is measured by the probability of rejecting H_0 when H_1 is true. This probability, denoted by P, is expressed by the relationship

$$P = 1 - \beta \tag{3.42}$$

It should be noted that fixing α determines the critical value of the test.

Tables 3.1 and 3.2 give critical values of r_c and d_c for the test statistics, and powers of both the considered test. The values were obtained by simulation using 20 000 trials of size N, for $N = 10$, 20, 30, 40 and 50.

These Tables enable one to select the size N of a sample in order to ensure satisfactory error probability and power P.

It is of interest to compare the results with those obtained by the test statistic based on the RML [13]. Both tests described here are only slightly inferior to the RML test. In particular, for sample size $N = 50$ the powers of the r test and RML test are equal.

The results obtained by Monte Carlo simulation indicate that using these classification tests (3.39) and (3.41), the log-normal and Weibull distributions can be fairly well identified. In the context of signal detection, however, it is of interest to determine the resulting improvement in the overall CFAR performance of a signal detector which uses such a classification test.

To illustrate the application of the simple classification test (3.41) to the log t detection procedure, we consider a signal detector (called the adaptive log t detector) which implements an algorithm shown in Fig. 3.9.

Table 3.1 *Critical values and power of the r test (from Szajnowski [13], © 1977 IEEE)*

	$\alpha = 0{\cdot}20$		$\alpha = 0{\cdot}10$		$\alpha = 0{\cdot}05$		$\alpha = 0{\cdot}01$	
N	r_c	P	r_c	P	r_c	P	r_c	P
10	−1·880	0·68	−1·795	0·54	−1·730	0·42	−1·630	0·09
20	−1·995	0·74	−1·880	0·59	−1·820	0·45	−1·725	0·24
30	−1·990	0·85	−1·910	0·73	−1·870	0·62	−1·780	0·38
40	−2·016	0·92	−1·950	0·81	−1·900	0·74	−1·820	0·51
50	−2·043	0·96	−1·980	0·91	−1·930	0·83	−1·850	0·63

Table 3.2 *Critical values and power of the d test (from Szajnowski [13], © 1977 IEEE)*

	$\alpha = 0{\cdot}20$		$\alpha = 0{\cdot}10$		$\alpha = 0{\cdot}05$		$\alpha = 0{\cdot}01$	
N	d_c	P	d_c	P	d_c	P	d_c	P
10	0·480	0·50	0·746	0·34	0·965	0·23	1·420	0·09
20	0·383	0·72	0·593	0·56	0·780	0·43	1·170	0·20
30	0·332	0·84	0·512	0·71	0·665	0·59	0·980	0·35
40	0·303	0·90	0·462	0·81	0·603	0·69	0·890	0·45
50	0·260	0·95	0·405	0·89	0·540	0·80	0·790	0·57

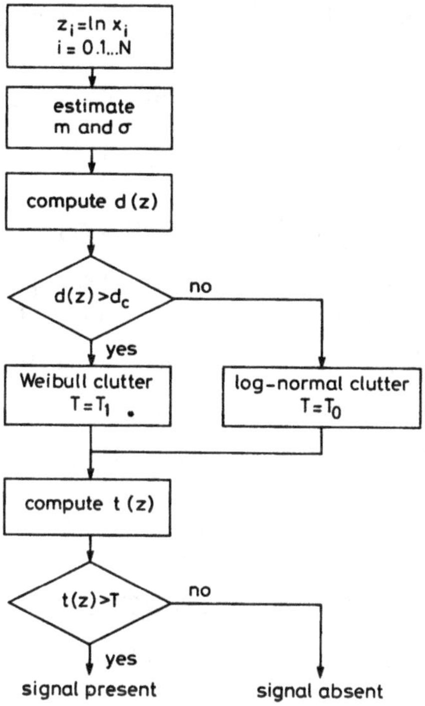

Fig. 3.9 *Flowchart for adaptive log t detection procedure (from Szajnowski [13], © 1977 IEEE)*

The adaptive log t detector makes the two decisions, namely:

(*a*) The local decision, based on eqn. 3.41 whether the clutter distribution is most likely log-normal or Weibull (this decision is then used to choose an appropriate detection threshold, T_0 or T_1);

(*b*) The global decision, based on the log t test

$$t(z) = (z_c - \hat{m})/\hat{\sigma} \tag{3.43}$$

where z_o denotes a sample from the resolution cell under test, to determine whether a signal is present.

Let

$$P_{fa}/\mathrm{Wbl}(T) = P_r\{t > T|\ \text{Weibull clutter only}\} \tag{3.44}$$

$$P_{fa}/\log(T) = P_r\{t > T|\ \text{log-normal clutter only}\} \tag{3.45}$$

$$P_d/\mathrm{Wbl}(T) = P_r\{t > T|\ \text{signal} + \text{Weibull clutter}\} \tag{3.46}$$

$$P_d/\log(T) = P_r\{t > T|\ \text{signal} + \text{log-normal clutter}\} \tag{3.47}$$

Thus, when only Weibull clutter is present, the overall false-alarm probability $\bar{P}_{fa}/\text{Wbl}$ can be expressed as

$$\bar{P}_{fa}/\text{Wbl} = PP_{fa}/\text{Wbl}(T_1) + (1-P)P_{fa}/\text{Wbl}(T_0) \tag{3.48}$$

Similarly, the overall false-alarm probability for log-normal clutter only, $\bar{P}_{fa}/\log$, can be obtained from

$$\bar{P}_{fa}/\log = (1-\alpha)P_{fa}/\log(T_0) + \alpha P_{fa}/\log(T_1) \tag{3.49}$$

For the overall detection probabilities in Weibull clutter and log-normal clutter, respectively, we have the following expression:

$$\bar{P}_d/\text{Wbl} = PP_d/\text{Wbl}(T_1) + (1-P)P_d/\text{Wbl}(T_0) \tag{3.50}$$

$$\bar{P}_d/\log = (1-\alpha)P_d/\log(T_0) + \alpha P_d/\log(T_1) \tag{3.51}$$

We shall now consider a simple example to demonstrate that, with the use of the adaptive log t detector, the overall CFAR performance can be satisfactory. If we choose $N = 50$ and $T = 2{\cdot}65$, using the Weibull threshold characteristics and percentiles of the Student's distribution, we obtain for the log t detector

$$P_{fa}/\text{Wbl}(T) \cong 10^{-4}, \quad \text{and} \quad P_{fa}/\log(T) \cong 5 \times 10^{-3} \tag{3.52}$$

It is seen that $P_{fa}/\log(T)$ differs from $P_{fa}/\log(T)$ by almost two orders of magnitude.

On the other hand, for the adaptive log t detector with $N = 50$, $T_1 = 2{\cdot}65$, $T_0 = 4{\cdot}2$, $d = 0{\cdot}405$ [hence $\alpha = 0{\cdot}1$ and $P = 0{\cdot}89$ (see Table 3.2)], we find from (3.48) and (3.49) that

$$P_{fa}/\text{Wbl} \cong 0{\cdot}89 \times 10^{-4}, \quad \text{and} \quad P_{fa}/\log \cong 5{\cdot}9 \times 10^{-3} \tag{3.53}$$

Note that, if the decision threshold d were equal to $0{\cdot}54$ ($\alpha = 0{\cdot}05$, $P = 0{\cdot}80$) the overall false-alarm probabilities would become

$$P_{fa}/\text{Wbl} \cong 0{\cdot}80 \times 10^{-4}, \quad \text{and} \quad P_{fa}/\log \cong 3{\cdot}45 \times 10^{-4} \tag{3.54}$$

By examination of eqns. 3.52, 3.53 and 3.54 it is evident that the CFAR performance of the adaptive log t detector can be superior to that of the original log t detector. However, the computation of test statistics $r(z)$, and even $d(z)$, needed for discrimination between log-normal and Weibull distributions, is too complex for real-time applications.

Another method of maintaining the false-alarm rate constant is to adjust the threshold adaptively. The threshold is calculated from the moments of the reference samples and the design value of the probability of false alarm P_{fa}. Tugnait and Prasad [15] suggested a CFAR detector with asymptotically regulated false-alarm rate.

In this method the PDF of clutter plus noise is approximated by a truncated generalised Laguerre polynomial expansion. The first five terms of the expansion are used.

Let $f(x)$ be the PDF of the clutter-plus-noise statistics. Then [16]

$$f(x) = \sum_{j=0}^{\infty} C_j \exp(-x/\beta)(x/\beta)^{\alpha} L_j^{\alpha}(x/\beta), \quad x \geqslant 0$$
$$= 0, \quad x < 0 \tag{3.55}$$

$L_j^{\alpha}(x)$ is a generalised Laguerre polynormial of order j, and α and β are chosen to make $C_1 = C_2 = 0$. The required values are $\alpha + 1 = [E(x)]^2/\mathrm{var}(x)$, $\beta = \mathrm{var}(x)/E(x)$. The coefficients of expansion C are as given in Ref. 16. The above choice of α and β makes the first two moments of the series equal to the first two moments, respectively, of $f(x)$.

Approximating $f(x)$ by terms until $j = 4$, and rearranging, we have

$$f_a(x) = \sum_{i=0}^{4} K_i x^{\alpha+i} \exp(-x/\beta), \quad x \geqslant 0$$
$$= 0, \quad x < 0 \tag{3.56}$$

where K_i are appropriately defined. The area under (3.56) is $A = \Sigma_{i=0}^{4} \beta^{i+1} K_i \Gamma(\alpha + i + 1)$. Note that $K_i \Gamma(\alpha + i + 1)$ does not depend on $\Gamma(\alpha)$, so that A can be computed without knowing $\Gamma(\alpha)$. After normalising (3.56) so that the area under $f_a(x)$ is unity, we still denote the normalised $f_a(x)$ by (3.56).

Four types of statistical models for radar clutter can be approximated by this truncated generalised Laguerre polynomial expansion: (i) Rayleigh, (ii) chi, (iii) log-normal, (iv) Weibull. The standard forms of these PDFs are

Weibull

$$f(x) = (\eta/\nu)(x/\nu)^{\eta-1} \exp[-(x/\nu)^{\eta}] \tag{3.57}$$

Chi:

$$f(x) = [2/\beta^{\alpha+1}\Gamma(\alpha+1)]x^{2\alpha+1} \exp(-x^2/\beta) \tag{3.58}$$

Log normal

$$f(x) = [\sqrt{(2\pi)}x\sigma]^{-1} \exp[-(\ln x - \ln \mu)^2/2\sigma^2] \tag{3.59}$$

The Rayleigh PDF is a special case of eqns. 3.57 and 3.58 obtained when $\eta = 2$ and $\alpha = 0$, respectively.

It is easy to see that the chi PDF at the square-law detector input is exactly described at the detector output by the first term of the Laguerre series, the coefficients of the rest of terms reducing to zero. This is also true for the Rayleigh PDF. Fig. 3.10*a* shows the actual and the approximation to the inverse distribution function for the Weibull PDF at the detector output with parameters $\eta = 0{\cdot}5$, $\nu = 1$. The first five terms of the Laguerre series were used for the approximation. $\nu = 1$ was chosen for convenience, with no loss of generality as it is only a scaling factor. Fig. 3.10*b* shows the actual and the approximation to the inverse distribution function for the log-normal PDF

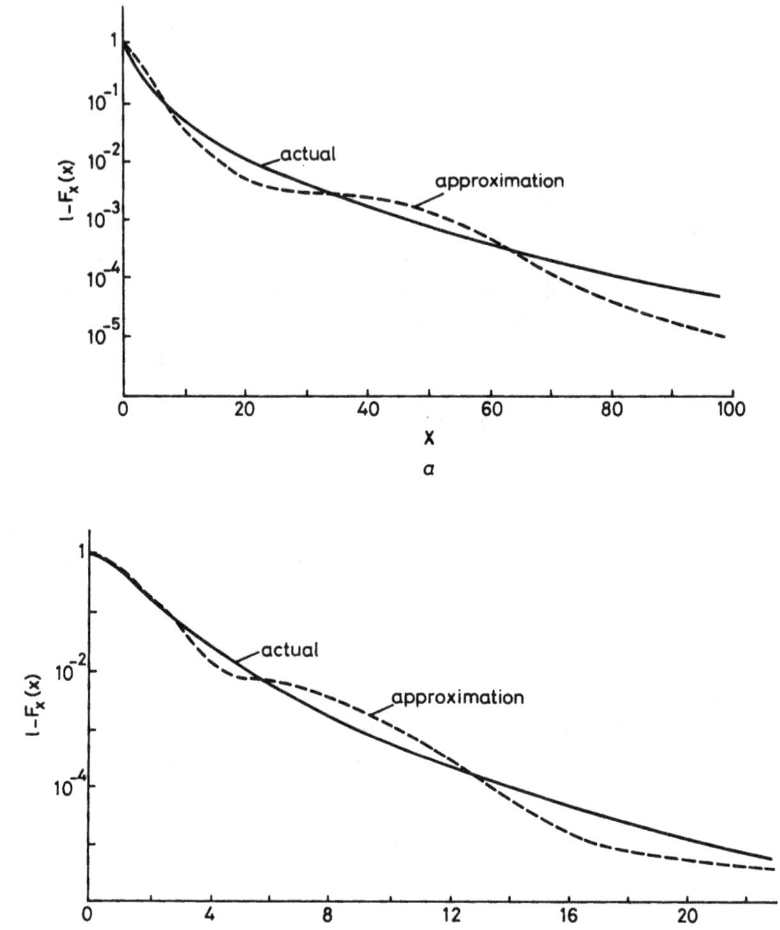

Fig. 3.10 *Laguerre series approximation to Weibull PDF (a), and log-normal PDF (b) (from Tugnait et al. [15], © 1977 IEEE)*

with parameters $\sigma = 0{\cdot}7147 = 3{\cdot}1075$ dB and $\mu = 1$. $\mu = 1$ was chosen for convenience, with no loss of generality as it is only a scaling factor.

Now the problem is how to calculate the adaptive threshold in terms of the clutter moments and the design value of the probability of false alarm P_{fa}. Use is made of the Chernoff (upper) bound on the resulting asymptotic expression for P_{fa} [17].

We have

$$P_{fa} = \int_{\gamma}^{\infty} f_{\text{approx}}(x)\, dx \tag{3.60}$$

where γ is the detection threshold. Since estimation of the clutter moments is assumed to be perfect, γ is a non-random (though unknown) number. Further

$$P_{fa} = \int_{\gamma}^{\infty} \exp[-sx + \mu(s)] f_{xs}(x)\, dx \tag{3.61}$$

where x_s, the tilted variate [Ref. 17, Sec. 2.7], has the PDF

$$f_{xs}(x) \triangleq \exp(sx) f_{\text{approx}}(x)/\exp[\mu(s)] \tag{3.62}$$

$$\mu(s) \triangleq \ln\{E[\exp(sx)]\} = \ln\left[\sum_{i=0}^{4} K_i' a^{-(\alpha+i+1)}\right] \tag{3.63}$$

$$\dot{\mu}(s) = E(x_s), \quad \ddot{\mu}(s) = \text{var}(x_s), \quad \dddot{\mu}(s) = E[x_s - E(x_s)]^3 \tag{3.64}$$

where $\dot{\mu}(s)$, $\ddot{\mu}(s)$ and $\dddot{\mu}(s)$ are the first three partial derivatives, respectively, of $\mu(s)$ with respect to s, and $a = \beta^{-1} - s$, $K_i' = K_i \Gamma(\alpha + i + 1)$. Note that in (3.63) the expectation of $\exp(sx)$ exists for the given $f_{\text{approx}}(x)$ only if $a > 0$, i.e. $s < \beta^{-1}$. Also, we do not need to know the value of $\Gamma(\alpha)$ to compute K_i'.

Eqn. 3.64 implies that $\mu(s)$ is a convex function of $s (s < \beta^{-1})$. Since $\dot{\mu}(s) > 0$ for all positive random variables, $\mu(s)$ is a monotone increasing function. So is $\dot{\mu}(s)$ as $\ddot{\mu}(s) \geqslant 0$ for all random variables.

To computer P_{fa} we approximate the PDF of x_s by a Gaussian PDF having mean and variance $\dot{\mu}(s)$ and $\ddot{\mu}(s)$, respectively. We are interested primarily in the case when a number (say n) of independent radar returns are processed at a time. In such a case the detector output is the sum of n independent and identically distributed random variables. Since the central limit theorem is most effective near the mean of the variate of interest, we first 'tilt' x such that $\dot{\mu}(s) = E(x_s) \sim \gamma$ [17] and then apply the central limit theorem. Then from (3.61)

$$P_{fa} \cong \exp[\mu(s) - s\dot{\mu}(s) + (s^2/2)\ddot{\mu}(s)]\ \text{erfc}_*\{[\gamma - \dot{\mu}(s)]/\sqrt{[\ddot{\mu}(s)]} + s\sqrt{[\ddot{\mu}(s)]}\} \tag{3.65}$$

where

$$\text{erfc}_*(d) \triangleq \int_{d}^{\infty} [\sqrt{(2\pi)}]^{-1} \exp(-x^2/2)\, dx$$

Using the approximation $\text{erfc}_*(x) \lessapprox [\sqrt{(2\pi)}]^{-1} \exp(-x^2/2)$ and noting that we want $\gamma \sim \dot{\mu}(s)$, we have

$$P_{fa} \cong \exp[\mu(s) - s\gamma](\exp\{-1/2[\gamma - \dot{\mu}(s)]^2/\ddot{\mu}(s)\}/[2\pi s^2 \ddot{\mu}(s)^{1/2}) \tag{3.66}$$

where $0 < s < \beta^{-1}$. ($s > 0$ since $s = 0$ implies no tilt and $s < 0$ implies tilt away from the threshold.)

Solving eqn. 3.66 for γ and choosing the solution which gives minimum

$|\gamma - \dot{\mu}(s)|$, we obtain

$$\gamma = \dot{\mu}(s) + s\ddot{\mu}(s)[-1 + (1 - 2[s\dot{\mu}(s) + \ln\{P_{fa}s\sqrt{[2\pi\ddot{\mu}(s)]}\} - \mu(s)/s^2\ddot{\mu}(s)])^{1/2}] \tag{3.67}$$

To find γ corresponding to a given P_{fa} we solve (3.67) iteratively. Since $0 < s < \beta^{-1}$ we first try $s = 1/2\beta$. If the computed γ is not in the vicinity of $\dot{\mu}(s)$, we choose $s = (\beta^{-1} - 1/2\beta)/2$ or $s = (1/2\beta - 0)/2$ according as $\gamma - \dot{\mu}(s) > 0$ or $\gamma - \dot{\mu}(s) < 0$, respectively, and so on until we get $|[\gamma - \dot{\mu}(s)]/\sqrt{[\ddot{\mu}(s)]}| \leqslant 0{\cdot}1$. This method converges since $\mu(s)$ and $\dot{\mu}(s)$ are monotone increasing functions of s, even though $\ddot{\mu}(s)$ may not necessarily be monotone increasing. The results are shown in Figs. 3.11*a* and *b* for Weibull and log-normal clutter, respectively.

Two sets of curves are shown. These correspond to $n = 1$ and $n = 4$, where n is the number of post-detection integrations. It can be seen from these curves that the 'bias' error, defined as the normalised difference between the design P_{fa} and the asymptotic P_{fa} corresponding to the computed threshold, lies within a fraction of an order of magnitude for $10^{-3} > P_{fa} > 10^{-8}$. The result for Weibull clutter is somewhat better than that for log-normal clutter. However, the calculation of the threshold is too complex, and is impossible to calculate in real time.

Cole and Chen [18] proposed a very simple unique doubly adaptive CFAR detector with good CFAR performance in Weibull clutter even in dense target environments. This doubly adaptive CFAR detector is based on the use of an auxiliary parallel adaptive detector to regulate the threshold of the conventional main adaptive CFAR detector. The auxiliary adaptive detector has a lower threshold setting. The threshold crossing rate of the auxiliary detector, which depends on the clutter statistics, is used to adjust the base multiplier setting of the main detector. The block diagram of this doubly adaptive CFAR detector is shown in Fig. 3.12.

It utilises a low-threshold detector to sense the deviation of the statistics from Rayleigh characteristics and modify the high threshold used for false-alarm control. The output from a linear detector is A/D converted, and is then loaded into a shift register and clocked along the register. The outputs from the reference cells on both sides of the cell to be examined for detection are summed, and the resultant is divided by the total number of samples summed to provide a measurement of the mean. This mean is multiplied by a constant K_L to provide a threshold T_L to an amplitude comparator to determine whether the amplitude of the cell to be examined exceeds T_L. If it exceeds the threshold, a 1 is loaded into a low-level detector register of length n cells. If the threshold is not exceeded, a 0 is loaded. The sum 1s in the register is a measure of the false-alarm rate at the low threshold. This sum is used to address a lookup table to provide a modifying multipler K_M to change the apparent mean for the target detection channel. The mean is multiplied by K_M and then K_H to provide a threshold T for target detection. The factor K_H is

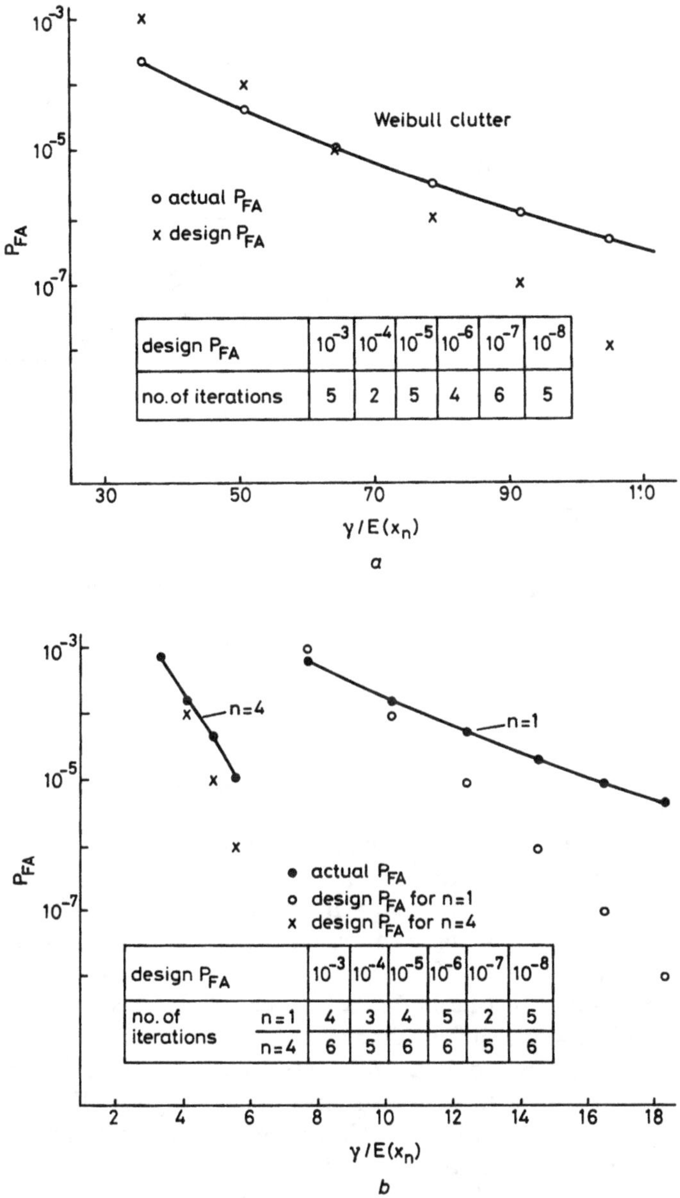

Fig. 3.11 *Actual and design P_{fa} versus normalised detection threshold for Weibull ($\eta = 0{\cdot}5$) (a), and log-normal ($\sigma = 3{\cdot}1075\,dB$) clutter (b) (from Tugnait et al. [15], © 1977 IEEE)*

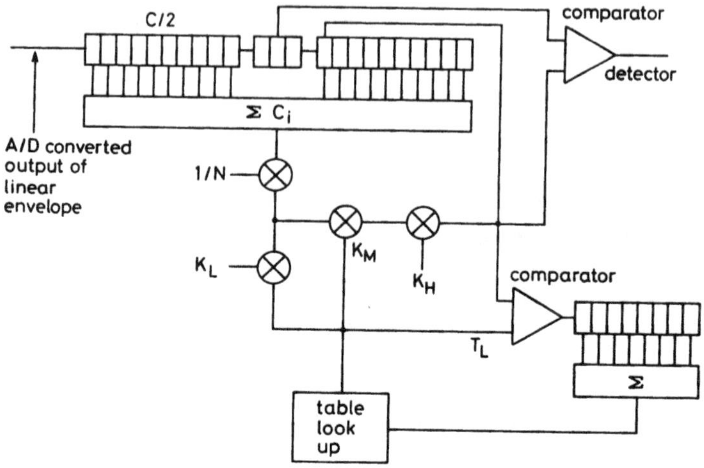

Fig. 3.12 *Block diagram of the doubly adaptive CFAR detector (from Cole et al. [18], © 1978 IEEE)*

picked to give the desired false-alarm rate for Rayleigh noise. The target-cell amplitude is compared with T to determine whether detection has occurred. The lookup table can be loaded in any desired fashion based on the class of statistics that are expected. As the number of detections in the low detector register increases, the value of K_M is increased, and vice versa, to control the high threshold.

The number of cells n must be substantial, e.g. 256, in order to provide a good measure of the statistics at the low threshold. On the other hand, n cannot be too large or there will be too much delay in modifying the threshold as the statistics vary. If n is too large, the probability of interference targets entering the reference cells will also increase. The multiplier K_L must be selected to give a substantial variation in false-alarm rate as the statistics vary. For Weibull clutter, a value of 2 appears to be a good choice and results in the virtual elimination of one multiplier in the process, as the multiplication becomes a simple one-bit shift in the mean. The two multipliers, K_M and K_H, can be combined in the lookup table, resulting in elimination of another multiplier. The CFAR performance of this doubly adaptive CFAR detector in Weibull clutter is shown in Fig. 3.13. These results were obtained by computer simulation using Monte Carlo techniques.

It can be seen from this Figure that, when the shape parameter α ($\alpha = 1/c$) of the Weibull distribution changes from 0·5 to 0·9 (i.e. c changes from 2·0 to 1·11) the false-alarm rate is maintained almost unchanged, especially for $P_{fa} > 10^{-6}$.

Some losses occur owing to the use of a finite number of clutter cells in the CFAR block and in the low-level detection register. The loss appears to be made up of two primary components which are additive. The first of these is

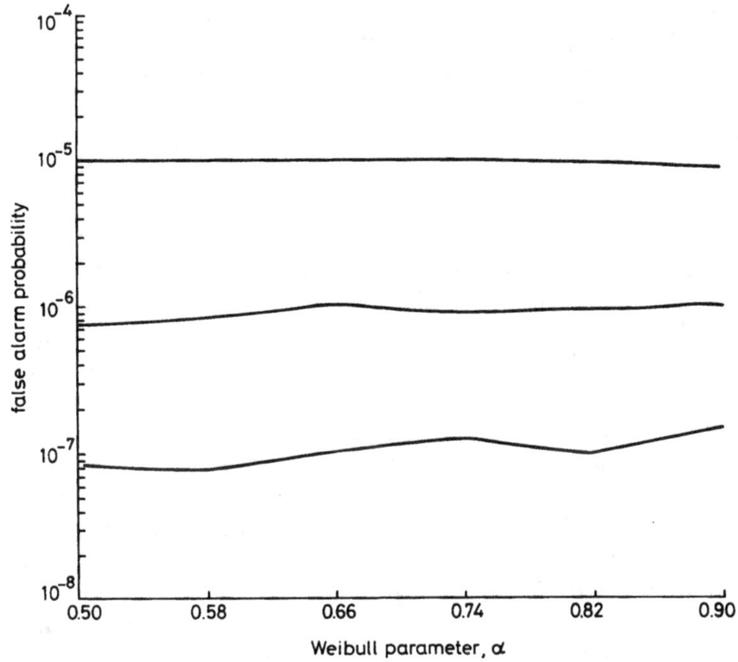

Fig. 3.13 *CFAR performance of the doubly adaptive CFAR detector (from Cole et al. [18], © 1978 IEEE)*

the sample-size loss in measuring the mean, and is essentially unchanged from the loss incurred in a standard mean-level detector. This loss is proportional to the false-alarm number and inversely proportional to the sample size. For a sample size of 128 and P_{fa} of 10^{-6}, the loss is about 0·2 dB. The second loss is due to the uncertainty in measuring the alarm rate at the low-level detector with a finite number of samples. It appears to follow the same laws of proportionality as the mean-level detector. The loss is about 1·1 dB for a register length of 512 at a P_{fa} of 10^{-6}. There is a slight variation in loss with α of about 0·5 dB as α varies from 0·5 to 0·8 (c varies from 2·0 to 1·25). The total loss is about 1·2 dB for $\alpha = 0{\cdot}5$ ($c = 2{\cdot}0$), 1·3 dB for $\alpha = 0{\cdot}63$ ($c = 1{\cdot}59$) and 1·7 dB for $\alpha = 0{\cdot}83$ ($c = 1{\cdot}21$) for the parameters quoted. These results were achieved with a 13-step lookup table. However, the sample size of 128 or 512 is too large for many practical cases; therefore, the loss will be greater than in this example.

This CFAR technique has other advantages. The first is that the required dynamic range of the linear receiver is rather smaller than that of the Goldstein's adaptive CFAR detector. No logarithmic detector is required; it requires only a linear detector with a dynamic range of 30 dB. On the other hand, it is difficult to achieve zero DC bias in the envelope detector and A/D converter. Any DC bias will result in either an increase or decrease in the

conventional CFAR technique. This process will sense the change, and tend to compensate for and minimise the effects of any residual DC bias.

It should be noted that the doubly adaptive detector can easily be extended to the triply or higher-order adaptive detector. The higher-order adaptive detector will undoubtedly provide more statistical information about the clutter, and therefore provide more precise control on the base-multiplier value.

There is a further problem for the conventional adaptive detector which fails to accommodate the target-to-target interference as targets pass through the clutter cells in the detector. The clutter cells are usually referred to as a CFAR block which forms the clutter average. When the target signal enters the CFAR block, this signal will raise the clutter average and degrade the ability to detect the nearby target incidentally located in the detection cell. The detection degradation due to the target interference has been found unacceptable, especially if radar operates in very heavy target environments. Fig. 3.14 shows the detection degradation as a function of the signal-to-clutter ratio, and the percentage of the CFAR block being occupied by target signals.

It can be seen from this Figure, that a target with signal-to-clutter ratio of 20 dB will yield 1·5 dB detection degradation if a target occupies only 2% of a CFAR block. The degradation is found to be even more severe if the target range extent or amplitude is large. Multiple targets such as a fleet of naval vessels or aircraft are not unusual in the real radar environment.

Cole and Chen [18] suggested a target discrimination technique to solve this

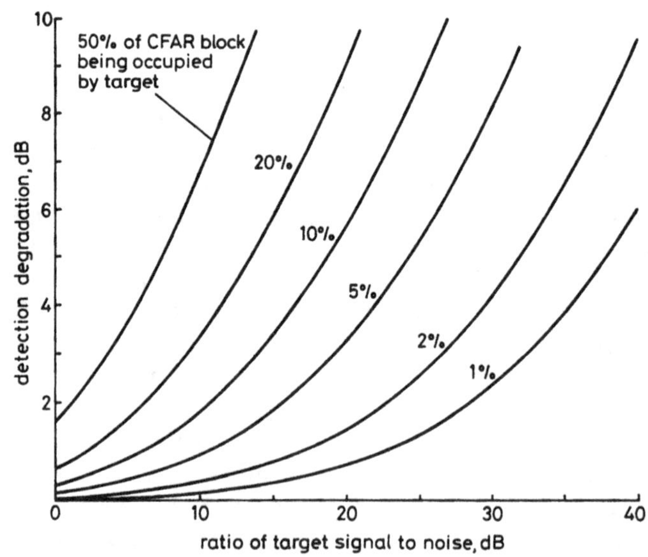

Fig. 3.14 *Detection degradation due to target interference (from Cole et al. [18], © 1978 IEEE)*

problem. This technique is based on the use of a simple logic circuit inserted between the detection cell and the clutter average cells or CFAR block. This logic circuitry prevents the target signal from entering the CFAR block by simply replacing the target signal with the clutter average established previously. Implementation of the target discrimination circuit in the adaptive detector is shown in Fig. 3.15; only a single-sided CFAR block will be considered.

The basic process of achieving the target discrimination technique is to insert a gated switch between the detection window and the CFAR block. The gated switch is controlled by the complement of the target detection.

If the signal amplitude in the detection window exceeds the threshold T, a detection is indicated or a 1 is loaded in the output of the detection comparator. This 1 triggers the inverter and then switches off the gated switch. In the next clock period, this will prevent the target signal in the detection window from moving into the next register located in the CFAR block. The location in the next register, which would have contained the target, could be filled by the previously established clutter average.

Once the target process passes through the detection cell, the output of the detection comparator returns to zero, which will energise the inverter and turn on the gated switch, and therefore resume normal CFAR operation.

With this technique, the estimated clutter average will not be raised by the interference target and no detection degradation will result.

In many practical cases where there can be a rapid change in clutter level, such as at a land/sea boundary or chaff boundary, the one-sided CFAR block is not practical. The target discrimination circuit can and has been incorporated in a CFAR detector with both leading and trailing blocks. In this case, one target can still suppress another target which is at a closer range, but will not suppress a target which is at a longer range.

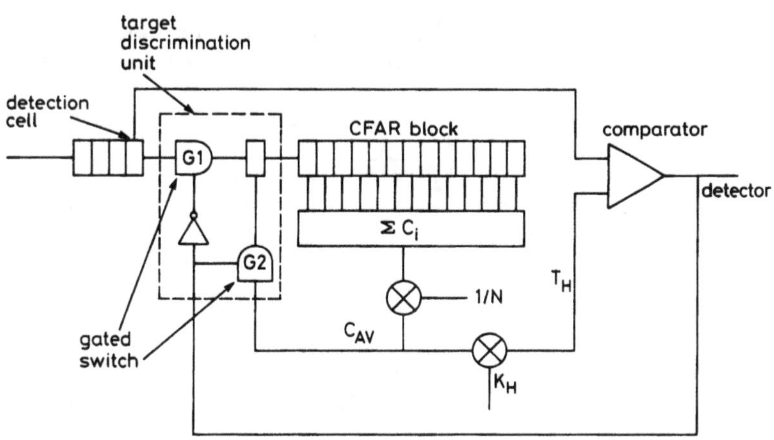

Fig. 3.15 *Implementation of target discrimination technique (from Cole et al. [18], © 1978 IEEE)*

The target discrimination technique can also be used in conjunction with the doubly adaptive CFAR detector. In this case, it prevents targets from loading into the low-level register, and thus prevents them from being sensed as a part of the skewness of clutter.

The 'greatest of' CFAR detector is often used to solve the clutter boundary problem. A block diagram of this 'greatest of' cell average CFAR detector is shown in Fig. 3.16.

It takes the mean values of each side of the clutter blocks, and then selects the greatest one, multiplying by some constant K as the normalising factor. If the leading edge enters the left side of the clutter block, the threshold will be raised. If the trailing edge leaves the right side of the clutter block, the threshold will also be raised. So it can maintain the false-alarm constant even in the clutter boundary. However, it will reduce the sensitivity of detecting a target which is located outside the edge of the clutter.

Bucciarelli [19] analysed the performance of this type of CFAR detector in the Weibull clutter, and suggested a method of maintaining the false-alarm constant.

The sample PDF of Weibull clutter is

$$p(x) = \frac{c}{b}\left(\frac{x}{b}\right)^{c-1} \exp\left[-\left(\frac{x}{b}\right)^{c}\right] \quad x \geqslant 0 \tag{3.68}$$

whose distribution function is

$$P(x) = 1 - \exp[-(x/b)^c] \tag{3.69}$$

The false-alarm probability is

$$P_{fa} = \int_0^{\infty} p(x) P^N\left(\frac{x}{K}\right) dx \tag{3.70}$$

where N is number of range samples before and after the cell under test, and K is the gain which is used to multiply the highest-side sample to obtain the

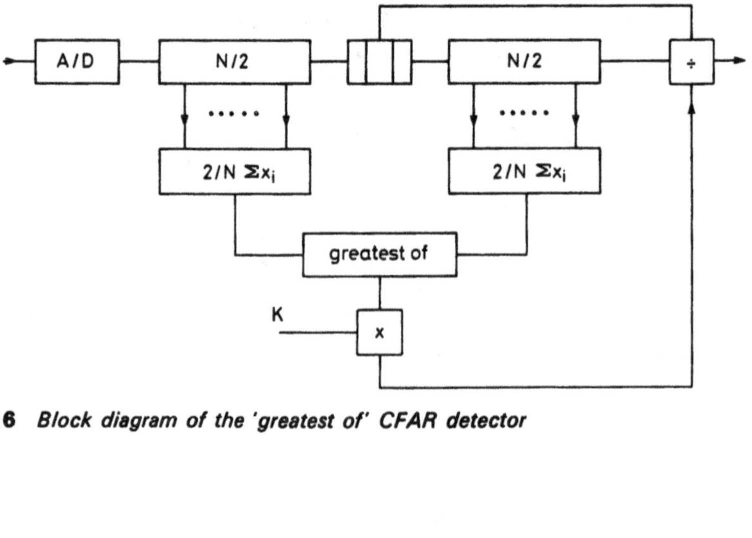

Fig. 3.16 *Block diagram of the 'greatest of' CFAR detector*

threshold. It is easy to verify that P_{fa} is independent of b (the Weibull scale factor), but not of c (its shape factor); so, for a fixed c, the CFAR can be obtained, but by changing c for a fixed K, different probabilities of false alarm are obtained.

As the clutter power is

$$W = 1/2(b)^2\Gamma(1+2/c) \tag{3.71}$$

the stated independence of b implies independence of power.

Eqn. 3.70 can be rewritten as

$$P_{fa} = \int_0^\infty \frac{c}{b}\left(\frac{x}{b}\right)^{c-1} \exp\left[-\left(\frac{x}{b}\right)^c\right]\left\{1-\exp\left[-\left(\frac{x}{Kb}\right)^c\right]\right\}^N dx \tag{3.72}$$

and after a few computations, the following relation is obtained:

$$P_{fa} = \sum_{h=0}^{N} \binom{N}{h} - \sigma(-1)^h \frac{K^c}{h+K^c} \tag{3.73}$$

This confirms that, for fixed c, a CFAR is obtained, the formula being very similar to that already known for Rayleigh clutter (i.e. $c = 2$).

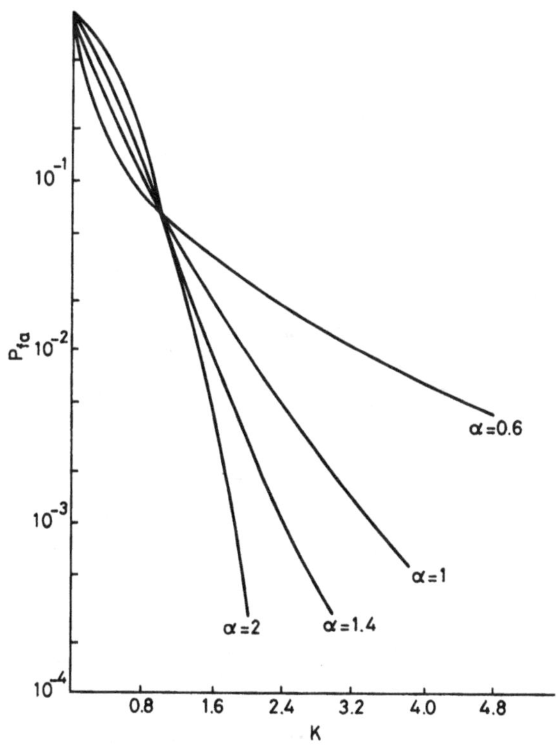

Fig. 3.17 *P_{fa} versus K for $N = 12$*

It is clear that, having chosen K to obtain the necessary P_{fa} under the Rayleigh assumption, if c changes then the CFAR is no longer possible; the smaller the value of c, the longer the distribution tails, and for a fixed K the higher the probability of false alarm.

In Fig. 3.17 the probability of false alarm P_{fa} is plotted against K for $N = 12$, with c changing from 0·6 to 2. As is already known, for $K = 1$ a value for P_{fa} is obtained which is not only independent of b, but also of c; this value of P_{fa} is

$$P_{fa} = \frac{1}{N+1} \tag{3.74}$$

which is usually too high for radar applications.

It is possible to define a variable to show how the CFAR has been obtained:

$$\gamma(c) = \frac{P_{fa}(c, K)}{P_{fa}(2, K)} \tag{3.75}$$

A good CFAR has been obtained when this value is near unity. In Fig. 3.18, $\gamma(1.4)$ and $\gamma(1)$ are shown against N for various values of P_{fa}. For $c = 2$ (the

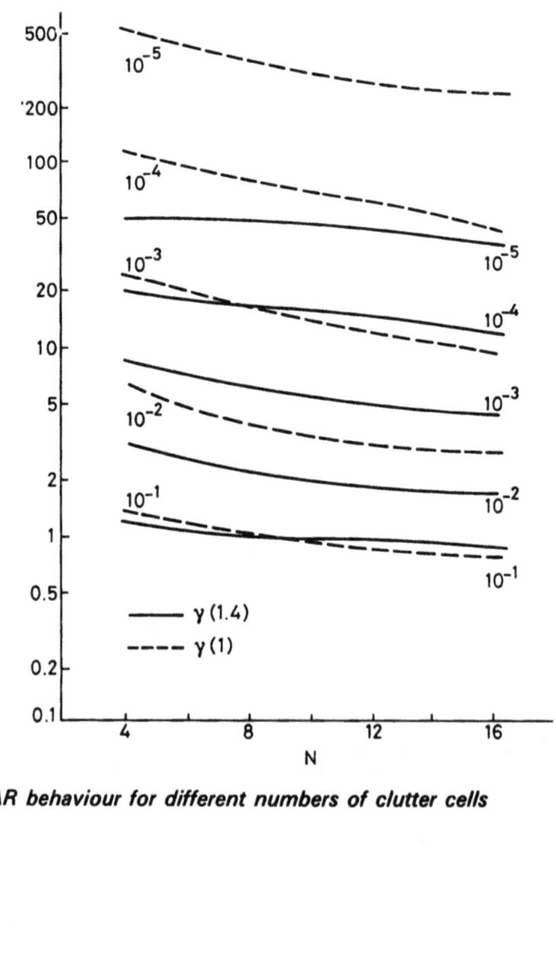

Fig. 3.18 *CFAR behaviour for different numbers of clutter cells*

Rayleigh case), values of K are computed which allow us to obtain the desired P_{fa}, from which the $\gamma(c)$ are evaluated.

For sea clutter c is assumed to vary from 1·4 to 2·0. It can be seen from this Figure that, if the number of cells is equal to 16, the desired P_{fa} is greater than 10^{-3} and $\gamma(1{\cdot}4)$ is less than 5. This means that, in this case, when the shape parameter changes from 2 to 1·4, P_{fa} only increases 5 times.

3.3 Non-parametric CFAR detector

All the CFAR detectors discussed in the previous Section belong to the parametric type of CFAR detector. This means that the distribution function is known *a priori* only the parameters of the distribution function has to be estimated. However, since the distributions of clutter are very complex, they depend not only on the type of the clutter, but also on the time of observation. Since most of the clutter is time-variable, the parameters and/or the type of the distribution may change from time to time. The CFAR detector designed for one type of distribution, such as Rayleigh or Weibull, cannot maintain the false-alarm rate constant for another type of distribution, or it can maintain the false-alarm rate constant sometimes but not always. In other words, the parametric CFAR detector cannot maintain the false-alarm rate constant in a real clutter environment. Therefore, the best way to solve this problem is to design a CFAR detector whose CFAR performance is distribution-free. This is the non-parametric CFAR detector.

Thomas [20] summarised the application of nonparametric statistical decision in signal detection. Dillard and Antoniak [21] first suggested a practical distribution-free detection procedure for multiple-range-bin radar, and they used a modified sign test to solve the distribution-free detection problem in radar techniques. Hansen and Olsen [22] proposed a generalised sign test for non-parametric radar detection. Trunk *et al.* [23] developed a modified generalised sign-test CFAR detector. This detector has been applied to a real radar system and produced good performance. The block diagram of a generalised-sign-test non-parametric CFAR detector is shown in Fig. 3.19.

The basic principle of this non-parametric CFAR detector is to evaluate the rank of the cell under test. Therefore, sometimes it is termed a rank detector. Let x_{ij} be the ith returned pulse in the jth range cell. The rank detector computes the rank R by making pairwise comparisons:

$$R_{ij} = \sum_{k}^{n} C(x_{ij} - x_{ik}) \tag{3.76}$$

where

$$C(x) = \begin{cases} 1, & x > 0 \quad \text{or} \quad x = 0; \quad ij \text{ odd} \\ 0, & x < 0 \quad \text{or} \quad x = 0; \quad ij \text{ even} \end{cases} \tag{3.77}$$

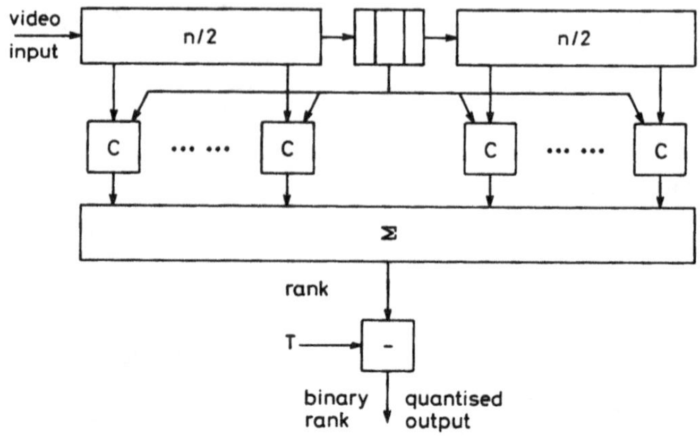

Fig. 3.19 *Block diagram of a non-parametric CFAR detector*

and the k summation is over the n range cells surrounding the jth cell. Therefore, the maximum value of the rank is equal to the number of reference cells.

There are two types of rank detector: the binary quantised rank detector, and the rank sum detector. Fig. 3.19 shows the binary quantised rank detector. In this type of rank detector, the rank of cell under test is compared with a threshold T, and a one-bit binary number is obtained at the output. Then it can be integrated by a binary moving-window integrator or two-pole filter integrator. In the rank sum detector, the rank of cell under test is a binary number more than one bit. It is sent to a multi-level moving-window detector or multi-level two-pole filter integrator directly.

It is very easy to prove that the non-parametric detector has a CFAR performance.

If the samples of reference cells are independent and of identical distribution, i.e. the IID assumption, the probability of the rank of the cell under test taking the value L is equal to

$$P(R = L) = 1/(n + 1) \tag{3.78}$$

where n is the number of reference cells.

This can be proved as follows.

If input samples $\{x\}$ satisfy the IID assumption, the PDF is denoted as $p(x)$. The probability of its rank R taking the value L is

$$P(R = L) = \binom{n}{L} \int_{-\infty}^{\infty} p(x)^{n-L}[1 - p(x)]^{L} \, dp(x), \quad \text{for } L = 0 \text{ to } n$$

If we denote $p(x) = t$, then

$$P(R = L) = \binom{n}{L} \int_0^1 t^{n-L}(1-t)^L \, dt$$

After changing the variable, we obtain

$$P(R = L) = \binom{n}{L} \int_0^1 (1-t)^{n-L} t^L \, dt$$

Since

$$p(x) = \sum_{L=0}^{n} P(R = L) x^L$$

therefore,

$$p(x) = \int_0^1 \sum_{L=0}^{n} \binom{n}{L} (1-t)^{n-L} (tx)^L \, dt$$

$$= \int_0^1 [(1-t) + tx]^n \, dt$$

$$= \int_0^1 [1 + t(x-1)]^n \, dt$$

Let $[1 + t(x-1)] = s$, so $ds = (x-1)\, dt$

$$p(x) = \frac{1}{x-1} \int_1^x s^n \, ds$$

$$= \frac{1}{x-1} \frac{1}{n+1} (x^n - 1)$$

$$= \frac{1}{n+1} \sum_{L=0}^{n} x^n$$

Therefore

$$P(R = L) = \frac{1}{n+1}$$

The probability of the rank taking the value equal to or greater than L is

$$P(R > = L) = 1 - L/(n+1) \tag{3.79}$$

Obviously, all these probabilities are independent of the distribution function of the samples of the reference cells. They only depend on the number of reference cells. Therefore, this type of detector is a distribution-free CFAR detector or non-parametric CFAR detector.

For a binary quantised-rank CFAR detector, there is an optimum threshold [24]. With this optimum threshold, the asymptotic loss is a minimum. When the number of reference cells is very large, the optimum threshold is equal to $0{\cdot}8(n+1)$, and the minimum asymptotic loss is equal to 0·94 dB. In this case, the probability of obtaining 1 from the quantiser, i.e. the rank-quantisation probability, which can be calculated from eqn. 3.79, is equal to 0·2. It is evident that this figure is too high for most practical cases. If there is a moving-window detector cascaded with this rank quantiser for video integration, the permitted rank-quantisation probability P_r can be calculated from [25]

$$P_{fa} = \binom{W-1}{M-1} P_r^M Q_r^{W-M+1} \tag{3.80}$$

where P_{fa} is the false-alarm probability at the output of the integrator, W is the length of the window and M is the decision criterion (second threshold) of the moving-window detector.

For example, if the window length $W = 16$, the optimum second threshold for Swerling 0 target $M = 8$, we can calculate the required rank-quantisation probability P_r from this formula, which is equal to 0·064 for a given $P_{fa} = 10^{-6}$. Since this value of P_r is very small, we often choose the threshold T of the quantiser $T = L = n$. Then the rank-quantisation probability P_r is equal to $1/(n+1)$. We can choose the number of reference cells n to satisfy the required rank-quantisation probability. In this example, n can be chosen to be 15, and then the final $P_{fa} = 0{\cdot}84 \times 10^{-6}$.

The values of P_{fa} for $n = 8$—20 and different M/W of practical interest are listed in Table 3.3.

Table 3.3 *P_{fa} versus n for different M/W*

	M/W			
n	5/8	6/12	8/16	9/20
8	$3{\cdot}7 \times 10^{-4}$	$3{\cdot}8 \times 10^{-4}$	$5{\cdot}2 \times 10^{-5}$	$4{\cdot}8 \times 10^{-5}$
10	$1{\cdot}5 \times 10^{-4}$	$1{\cdot}3 \times 10^{-4}$	$1{\cdot}3 \times 10^{-5}$	$1{\cdot}0 \times 10^{-5}$
12	$6{\cdot}8 \times 10^{-5}$	$5{\cdot}5 \times 10^{-5}$	$3{\cdot}8 \times 10^{-6}$	$2{\cdot}7 \times 10^{-6}$
14	$3{\cdot}5 \times 10^{-5}$	$2{\cdot}5 \times 10^{-5}$	$1{\cdot}4 \times 10^{-6}$	$8{\cdot}6 \times 10^{-7}$
16	$1{\cdot}9 \times 10^{-5}$	$1{\cdot}3 \times 10^{-5}$	$5{\cdot}4 \times 10^{-7}$	$3{\cdot}1 \times 10^{-7}$
18	$1{\cdot}1 \times 10^{-5}$	$6{\cdot}5 \times 10^{-6}$	$2{\cdot}3 \times 10^{-7}$	$1{\cdot}2 \times 10^{-7}$
20	$7{\cdot}0 \times 10^{-6}$	$3{\cdot}7 \times 10^{-6}$	$1{\cdot}1 \times 10^{-7}$	$5{\cdot}3 \times 10^{-8}$

The number of reference cells cannot be chosen to be too large. Since the clutter is non-homogeneous in the range direction, too many reference cells will cause the IID assumption to be out of order. It is hard to say how many reference cells constitutes the upper bound, since it depends on the absolute size of the resolution cell of the radar. In general, the smaller the size of the resolution cell, the more reference cells can be adopted.

In the case of the rank sum detector, the test statistic is

$$T_{Gs} = \sum_{j=1}^{M} R_j$$

where M is number of pulses to be integrated.

The probability of rank R taking the value $0 - n$ is equal to $1/(n+1)$, and is independent of the distribution of the input.

For the purposes of calculating the false-alarm rate of the rank sum detector, we have to find the probability-density function of the test statistic. It is well known that the probability-density function of the sum of M independent variables is equal to the convolution of these M probability density functions. We can obtain this by numerical calculation, an example being shown in Fig. 3.20.

In this Figure the number of reference cells is equal to 14, and the number of integrated pulses is equal to 15. The position of the maximum is 105. If given the threshold, the probability of false alarm can be calculated. It does not depend on the PDF of the input signal, but only on the number of reference cells and the number of pulses integrated.

Akimov [26] derived the false-alarm probability of the rank sum detector, which can be expressed as

$$P_{fa} = \left(\frac{1}{n+1}\right)^M \sum_{m=T}^{Mn} \sum_{t=0}^{m/n+1} (-1)^t \binom{M+m-t(n+1)-1}{M-1} \binom{M}{t} \qquad (3.81)$$

where n is the number of reference cells, M is the number of integrated pulses, and T is the threshold.

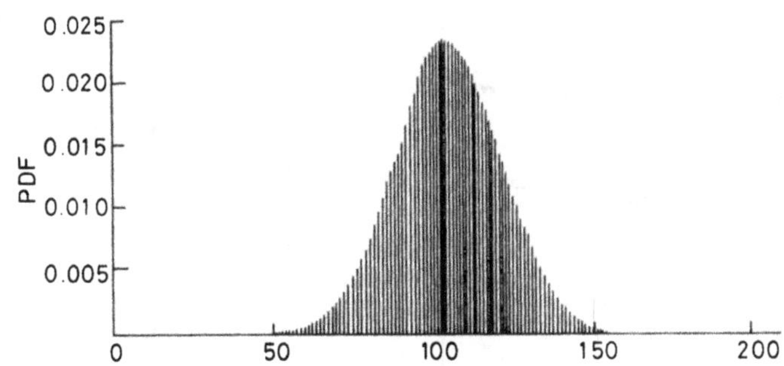

Fig. 3.20 *An example of the PDF of the test statistic of rank sum*

Table 3.4 *Required threshold T for $P_{fa} = 10^{-6}$*

M / n	8	9	10	11	12	13	14	15	16	17	18	19	20
8	62	68	75	81	87	93	97	104	110	116	121	127	132
10	77	85	92	100	108	115	122	129	136	143	150	157	164
12	91	101	110	119	128	137	146	154	163	171	179	188	196
14	106	117	128	139	149	159	169	179	189	199	208	218	229
16	121	133	146	158	170	181							

Table 3.5 *Required threshold T for $P_{fa} = 10^{-5}$*

M / n	8	9	10	11	12	13	14	15	16	17	18	19	20
8	60	66	72	78	84	90	95	101	106	112	117	122	128
10	74	82	90	97	104	111	118	125	132	138	145	152	158
12	89	98	107	115	124	132	141	149	157	165	173	181	189
14	103	114	124	134	144	154	163	173	182	192	201	210	220
16	117	129	141	153	164	175							

Table 3.6 *Required threshold T for $P_{fa} = 10^{-4}$*

M / *n*	8	9	10	11	12	13	14	15	16	17	18	19	20
8	58	63	69	75	80	86	91	96	102	107	112	117	122
10	71	78	86	92	99	106	113	119	126	132	139	145	152
12	85	94	102	110	118	126	134	142	150	158	166	173	181
14	99	109	118	128	137	147	156	165	174	184	193	201	210
16	112	124	135	146	157	167							

Table 3.7 *Required threshold T for $P_{fa} = 10^{-3}$*

M / *n*	8	9	10	11	12	13	14	15	16	17	18	19	20
8	54	59	65	70	75	81	86	91	96	101	106	111	115
10	67	74	80	87	93	100	106	112	119	125	131	137	143
12	80	88	96	104	111	119	127	134	142	149	157	164	171
14	93	102	111	120	129	138	147	156	165	173	182	191	199
16	105	116	127	137	147	158							

Table 3.8 *Required threshold T for $P_{fa} = 10^{-2}$*

M n	8	9	10	11	12	13	14	15	16	17	18	19	20
8	49	54	59	64	69	74	79	83	88	93	98	103	107
10	61	67	73	80	86	92	98	104	110	116	121	127	133
12	73	80	88	95	102	110	117	124	131	138	145	152	159
14	84	93	102	110	119	127	136	144	152	161	169	177	185
16	96	106	117	126	136	145							

Table 3.9 *Required threshold T for $P_{fa} = 10^{-1}$*

M n	8	9	10	11	12	13	14	15	16	17	18	19	20
8	42	46	51	56	60	64	69	73	78	82	86	91	95
10	52	58	63	69	75	80	86	91	97	102	108	113	119
12	62	69	74	82	89	96	102	109	116	122	129	135	142
14	72	80	88	96	104	112	119	127	134	142	150	158	165
16	82	91	100	109	118	127							

In practical cases, we are interested in how to determine the threshold for a given false-alarm probability. However, it is very difficult to calculate the threshold for a given P_{fa} from (3.81). Some numerical results are given in Table 3.4 for $P_{fa} = 10^{-6}$.

For example, it means that, if the number of reference cells is equal to 14, the number of pulses integrated is also equal to 15 just as in the case shown in Fig. 3.20. Then the required threshold should be equal to 179 to obtain the given $P_{fa} = 10^{-6}$.

Similar Tables could be calculated for other given false-alarm probabilities. Tables 3.5—3.9 list the results for $P_{fa} = 10^{-5}$—10^{-1}.

Once the threshold is set, the false-alarm probability is constant and does not vary with the input signal. This means that one can obtain any desired false-alarm probability by adjusting the threshold. Therefore this is the 'distribution free' CFAR feature of the non-parametric CFAR detector. The trade-off of this feature is the larger detection loss (or CFAR loss) which will be discussed later.

3.4 Signal detection in Weibull clutter

The theory of signal detection in Rayleigh noise was propounded by S. O. Rice, and further developed by Marcum and Swerling. The Rayleigh distribution can be used to describe the receiver noise very well. However, as mentioned in previous Sections, in most cases the distribution of clutter cannot be fitted with a Rayleigh function, but can be fitted with a Weibull distribution. Therefore, if we are interested in signal detection in clutter, we have to study the signal detection in Weibull-distributed background noise, but not in Rayleigh-distributed noise.

Ekstrom [5] first discussed the problem of signal detection in Weibull clutter theoretically. However, since it is very difficult to obtain a general closed-form expression for the PDF of signal plus Weibull clutter, he assumed that the signal vector is very much greater than the median value of the clutter. This is not true in many practical cases, and therefore the results are also meaningless.

When the Marcum–Swerling analysis is applied to the Weibull clutter situation, an immediate problem is encountered—the likelihood ratio cannot be found in closed form. This is a consequence of the unavailability of an analytical expression for the PDF under the signal-plus-clutter hypothesis.

Schleher [6] suggested a Weibull–Rician probability function to solve this problem.

By applying the procedure used by Rice to determine the Rician distribution, it can be shown that the normalised (median $V_m = 1$) density function of a steady signal with amplitude A in Weibull clutter is given by

$$f_v(v_s) = \int_0^{2\pi} [(c \ln 2V_s)/2]\{-\ln 2(V_s^2 - 2AV_s \cos\theta + A^2)^{c/2}]$$

$$/(V_s^2 - 2AV_s \cos\theta + A)^{(2-c)/2}\}\, d\theta \quad (3.82)$$

where c is the shape parameter of Weibull distribution.

Note that this distribution reduces to the Weibull distribution for $A = 0$. Since the Weibull distribution is a two-parameter distribution, even the normalised Weibull–Rician distribution is a function of c. For a given c, we can obtain one set of curves for different signal amplitude A. Fig. 3.21 shows the Weibull–Rician density function for $c = 1{\cdot}2$ and $A = 0$ to 8. Note that the density function is highly concentrated in the vicinity of A.

With this probability-density function for the signal-plus-clutter hypothesis Schleher calculated the detection performance for different types of receivers, including the linear receiver, logarithmic receiver, binary integrator and median detector.

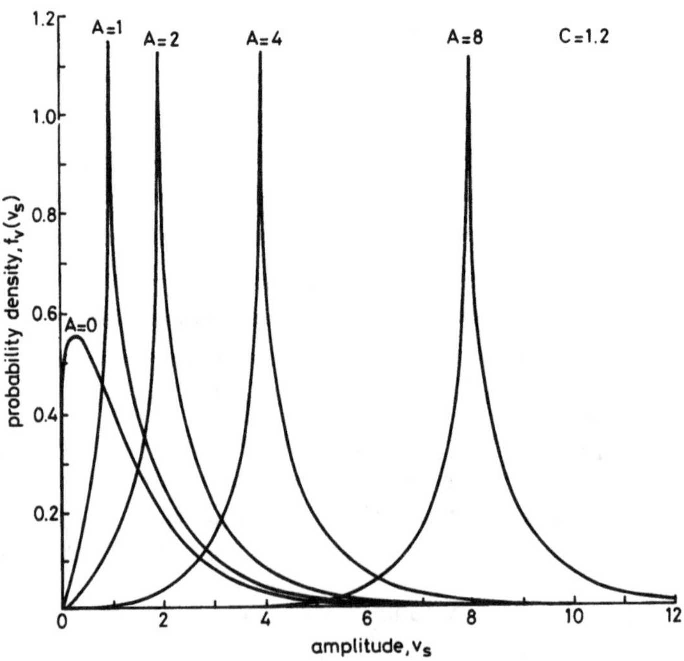

Fig. 3.21 *Weibull–Rician probability-density function for c = 1·2 (from Schleher [6], © 1976 IEEE)*

3.4.1 Detection performance of linear receiver in Weibull clutter

The linear receiver (linear envelope detector with linear integrator) is known to approximate to the optimum reciever in Rayleigh-distributed clutter. Since many radars are designed with linear receivers, it is important to determine how these receivers perform in Weibull clutter.

The performance of the linear receiver is analysed using a Marcum–Swerling analysis, where the Weibull distribution and Weibull–Rician distribution are used instead of Rayleigh and Rician distributions for hypotheses H_0 and H_1, respectively. The thresholds as a function of P_{fa} are determined for a number of independent samples $n = 1$, 3, 10 and 30 using a characteristic-function approach in conjunction with the FFT. P_d is determined in a similar manner using the FFT in conjunction with Gaussian quadrature to evaluate the sum distributions.

The results of the linear-receiver performance evaluation for $P_{fa} = 10^{-6}$ are shown in Fig. 3.22 for $c = 0{\cdot}6$, 0·8, 1·2 and 2·0 and $n = 1$, 3, 10 and 30. However, it should be noted that the number of independent samples n is somewhat different from the number of pulses within the beamwidth to be integrated. The reason is that the echoes from the target, and even from the clutter are not independent samples. In the target case, if and only if the fluctuation of echoes belongs to Swerling cases II and IV, i.e. pulse-to-pulse fluctuation, all the echoes can be seen as independent samples. In the clutter case, since the clutter echoes are correlated within the resolution cell, they cannot be treated as receiver noise which is uncorrelated between adjacent sweeps. Therefore, these results cannot be used in practical cases, but can only be used for relative performance comparison. However, they are valid in the case of frequency-agility radar. In this case, all the echoes are independent owing to the decorrelation effect of frequency agility, if the difference of carrier frequency between pulses is greater than 'critical frequency'.

Table 3.10 gives the additional signal-to-median clutter ratio required for the performance ($P_d = 0{\cdot}9$, $P_{fa} = 10^{-6}$) in Weibull clutter ($c = 0{\cdot}6$, 0·8 and 1·2) to equal that in Rayleigh clutter ($c = 2{\cdot}0$).

Examination of Table 3.10 shows that performance degradation is severe for the higher skewed Weibull clutter distributions. However, comparison with the linear-receiver performance obtained in log-normal clutter [27] shows that less degradation occurs in Weibull clutter.

3.4.2 Detection performance of a logarithmic receiver in Weibull clutter

A logarithmic detector has an output voltage whose amplitude is proportional to the logarithm of the input envelope. This type of detector, combined with the cell-averaging CFAR detector, has found extensive use in Rayleigh-distributed clutter owing to the constant false-alarm-rate performance.

The probability-density function of the logarithmic-detector output is different from that of the linear detector for the same input. In the Weibull-clutter

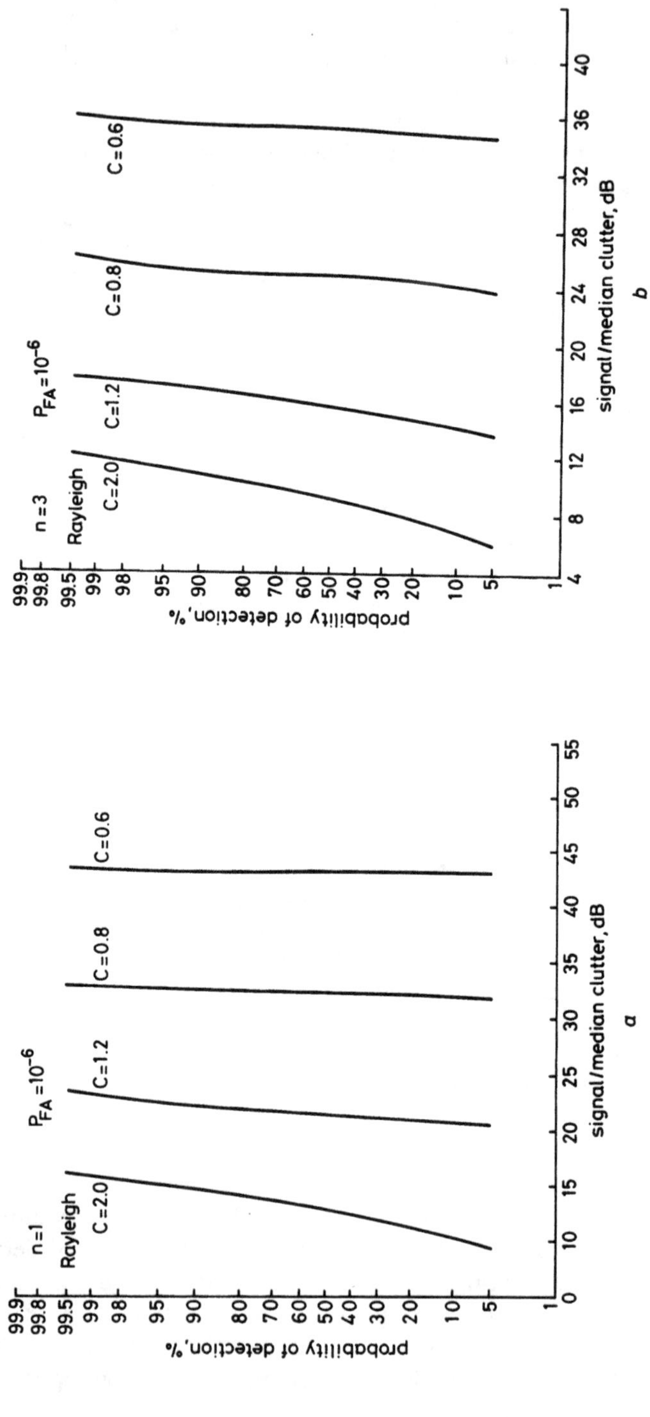
n=3
Rayleigh
C=2.0
$P_{FA}=10^{-6}$
C=1.2
C=0.8
C=0.6
probability of detection, %
signal/median clutter, dB
b
n=1
Rayleigh
C=2.0
$P_{FA}=10^{-6}$
C=1.2
C=0.8
C=0.6
probability of detection, %
signal/median clutter, dB
a

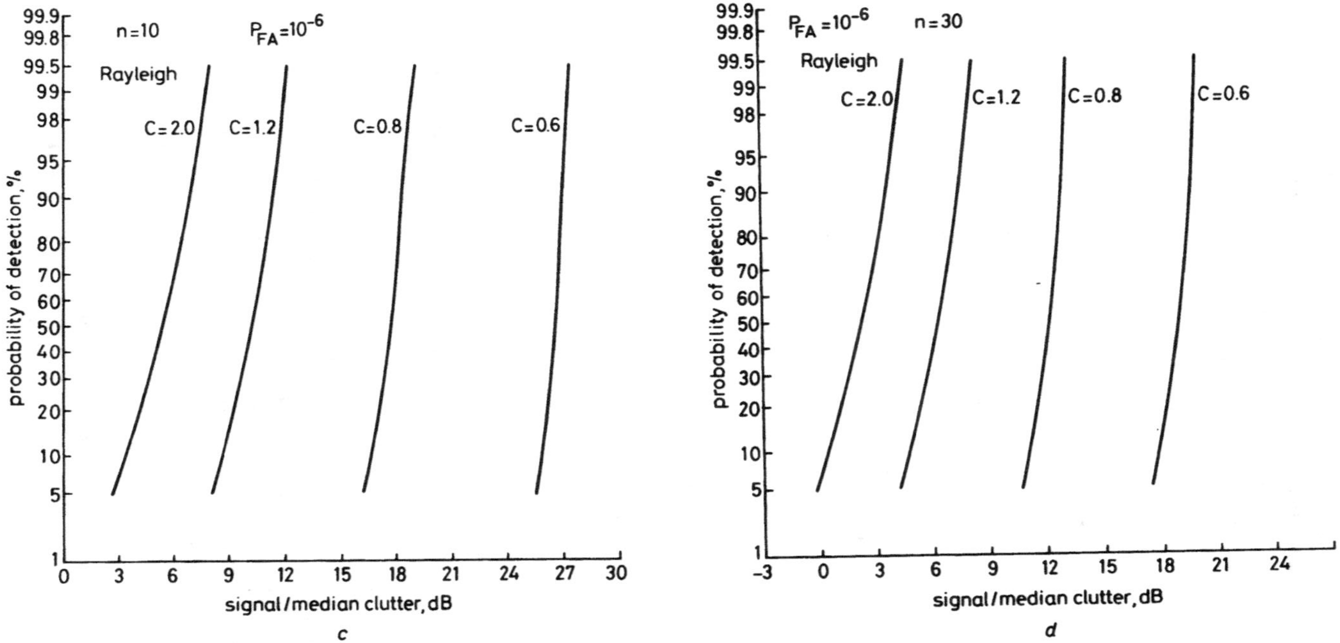

Fig. 3.22 *Detection performance of linear envelope detector in Weibull clutter (from Schleher [6], © 1976 IEEE)*

Table 3.10 *Linear receiver performance in Weibull clutter compared with Rayleigh clutter (additional signal-to-median clutter ratio in dB for $P_d = 0{\cdot}9$, and $P_{fa} = 10^{-6}$) (from Schleher [6], © 1976 IEEE)*

	c		
n	1·2	0·8	0·6
1	7·5	17·7	28·3
3	6·1	14·4	24·7
10	4·4	11·5	20·0
30	3·9	9·3	16·0

case, the PDF of the output of the logarithmic detector is given by

$$f_l(V_c \mid H_0) = c \ln 2 \exp(cV_c) \exp[-\ln 2 \exp(cV_c)] \tag{3.83}$$

Unde the signal-in-clutter hypothesis H_1, the PDF is given by

$$f_l(V_s \mid H_1) = \int_0^{2\pi} [c \ln 2 \exp(2V_s)/2\pi](\exp\{-\ln 2[\exp(2V_s) - 2A \exp(V_s) \cos\theta + A^2]^{c/2}\}/[\exp(2V_s) - 2A \exp(V_s) \cos\theta + A^2]^{(2-c)/2}) \, d \tag{3.84}$$

where the variable V_s and V_c extend from $-\infty$ to $+\infty$.

Fig. 3.23 shows the performance of the logarithmic receiver in Weibull clutter ($c = 0{\cdot}5, 0{\cdot}6, 0{\cdot}8, 1{\cdot}2, 2{\cdot}0$) for $aP_{fa} = 10^{-6}$ and $n = 1$, 3, 10 and 30 (independent samples). A comparison of the signal-to-median clutter differential for equal probabilities of detection ($P_d = 0{\cdot}9$, $P_{fa} = 10^{-6}$) in Weibull and Rayleigh clutter is given in Table 3.11.

Table 3.11 *Logarithmic receiver performance in Weibull clutter compared with Rayleigh clutter (additional signal-to-median clutter ratio in dB for $P_d = 0{\cdot}9$, $P_{fa} = 10^{-6}$) (from Schleher [6], © 1976 IEEE)*

n	1·2	0·8	0·6	0·5
1	7·5	17·8	28·3	37
3	6·0	12·6	20·1	26·0
10	3·6	7·8	12·2	15·9
30	2·4	3·8	6·1	7·9

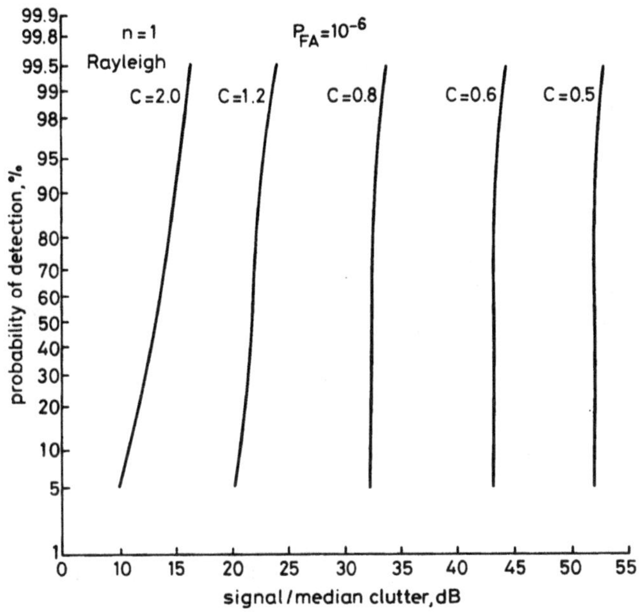

a

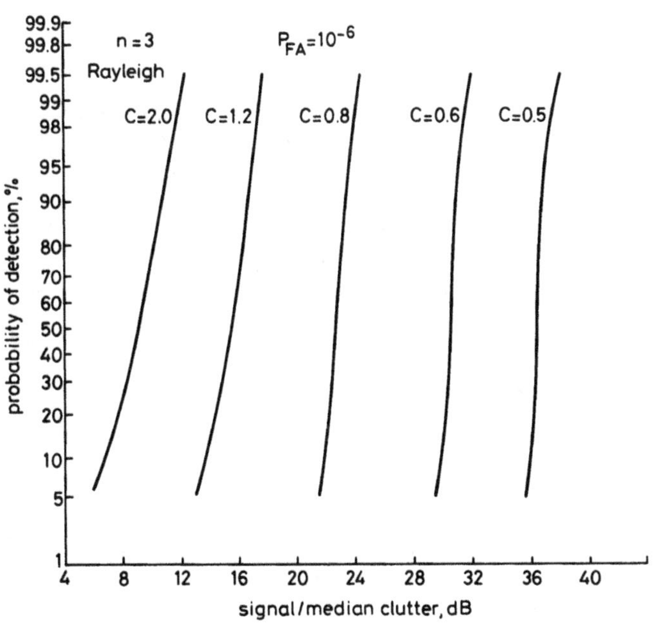

b

Fig. 3.23

(Continued overleaf)

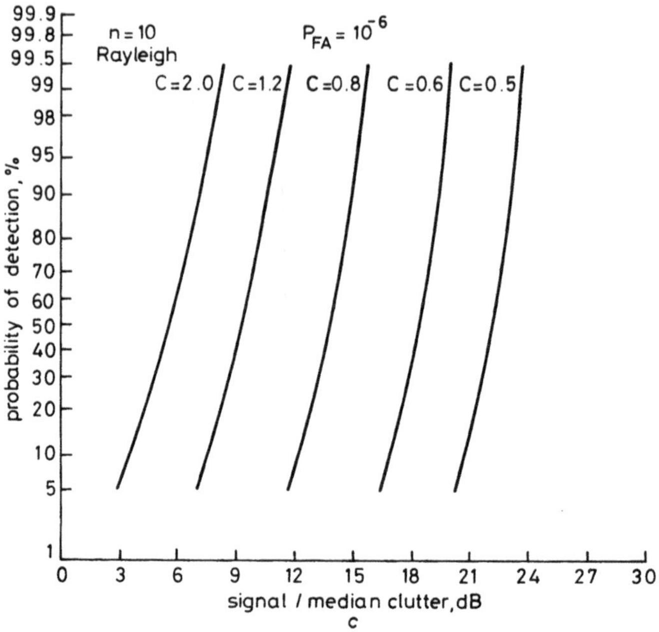

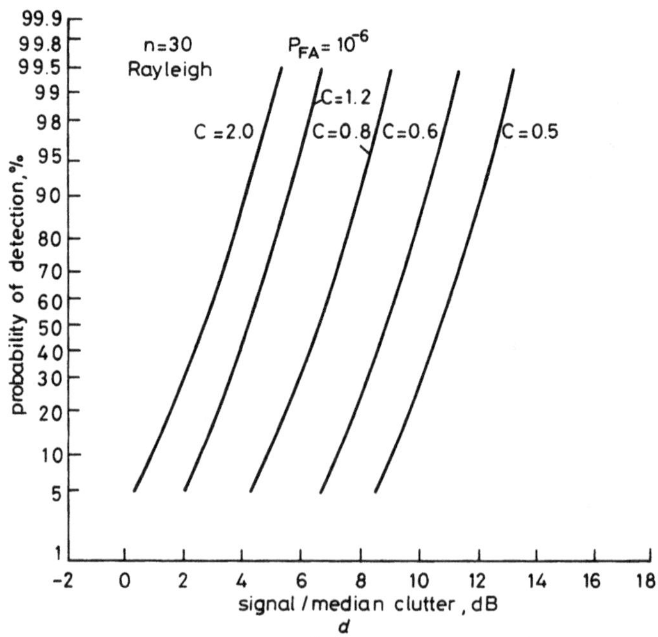

Fig. 3.23 *Detection performance of logarithmic receiver in Weibull clutter (from Schleher [6], © 1976 IEEE)*

3.4.3 Detection performance of binary integrator in Weibull clutter

The binary integrator, depicted in Fig. 3.24, is a digital detection process that utilises a double threshold.

The envelope-detected signal is compared with the first threshold. The number of threshold crossings in n repetitions of signal is counted. When more than m crossings take place in the n trials, a target is assumed to be present.

It should be noted that this double-threshold binary integrator is different from the practically used moving-window detector. The double-threshold integrator belongs to block processing, while the moving-window integrator belongs to moving processing. Since the azimuth position of a target is unknown *a priori*, the moving-window integrator is preferred in practical radar system.

It was determined by Schwartz [28] that the optimum second threshold of the binary integrator for a Rayleigh-distributed envelope-detected background should be set at approximately $1{\cdot}5\sqrt{n}$ for minimum required S/N ratio. This result is true for a broad range of false-alarm probabilities (10^{-5} to 10^{-10}) and probabilities of detection (0·5 to 0·9). The performance of this suboptimum detector in Rayleigh clutter is within 1·5 dB of the optimum performance.

The performance of the binary integrator in Weibull clutter was determined using the following procedure. First, using the Weibull-distribution function find the threshold setting n for all combinations of m and n of interest for a particular P_{fa}. Secondly, determine for each m and n of interest the cumulative probability from the envelope detector that provides the desired P_d. Then, using a graphical plot of the single-sample Weibull–Rician distribution function, determine the signal value that corresponds to the cumulative probability for the determined threshold setting n.

Using the preceding procedure, detection curves for the binary integrator were formed for $c = 1{\cdot}2$ over the region of interest, and the second threshold setting for the best performance was identified. The optimum values of m so found are given in Table 3.12 and compared with the optimum value in Rayleigh clutter ($1{\cdot}5\sqrt{n}$).

The detection performance of the binary integrator is shown in Fig. 3.25 and compared with the optimum performance in Weibull clutter. It can be seen from these Figures that the performance of binary integrator is very close to the optimum performance.

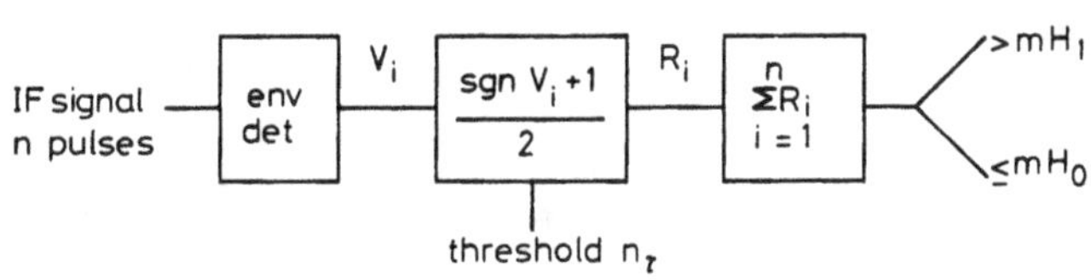

Fig. 3.24 *Block diagram of binary integrator (from Schleher [6], © 1976 IEEE)*

Table 3.12 *Optimum value of second threshold, Weibull clutter ($c = 1{\cdot}2$) (from Schleher [6], © 1976 IEEE)*

n	m_{opt}	m
3	2	2
10	7	4
30	23	8

3.4.4 Detection performance of median detector in Weibull clutter

The median detector can be implemented by employing the binary integrator with the second threshold set at $m = (n - 1)/2$. This effectively accomplishes the process of finding the median of the input distribution and then comparing it against a threshold.

The median detector can be analysed using a procedure identical to that used for the binary integrator. Detection curves were formed for $c = 1{\cdot}2$ and are shown in Fig. 3.25. The performance is inferior to other detectors

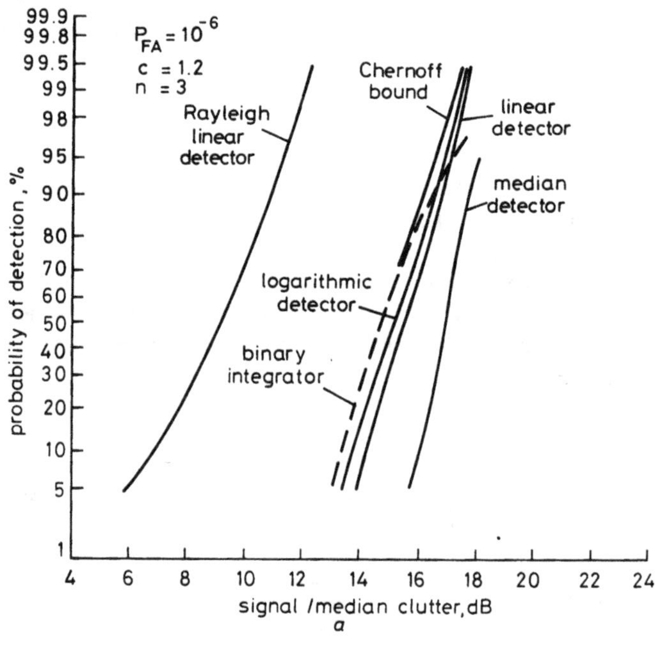

Fig. 3.25

(Continued overleaf)

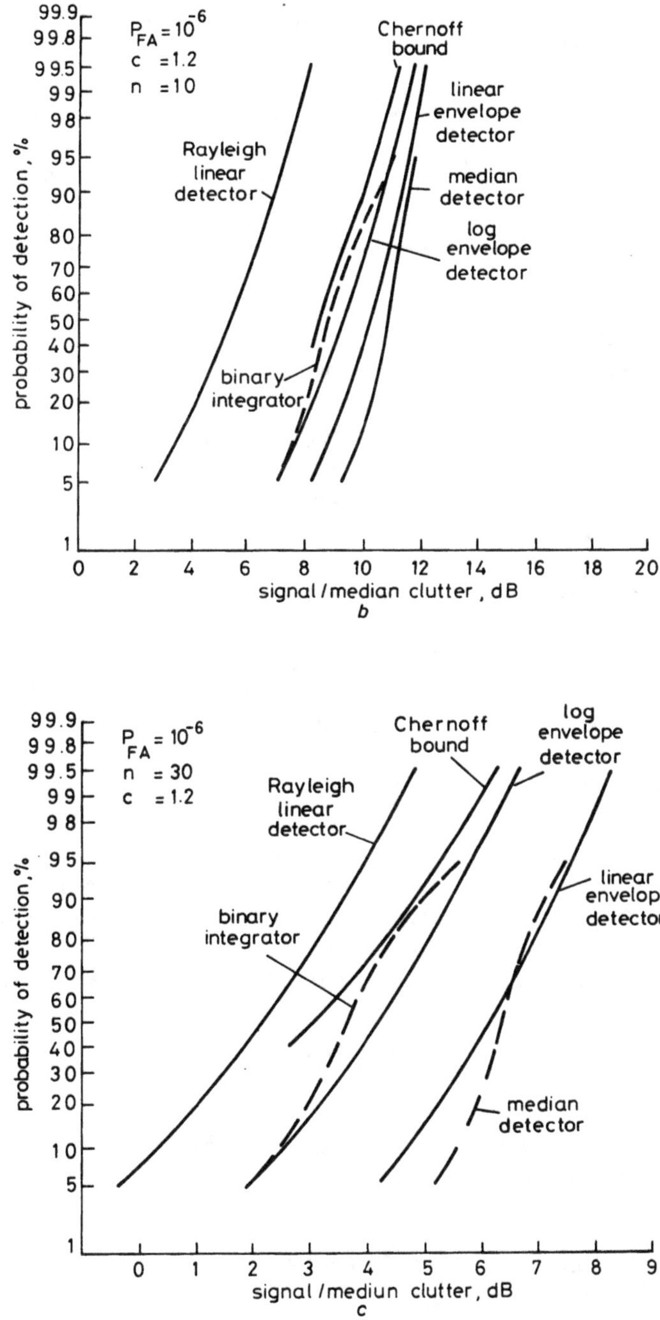

Fig. 3.25 *Comparison of detection performance, Weibull clutter (from Schleher [6], © 1976 IEEE)*

Table 3.13 *Signal-to-median clutter in dB for median detector ($P_{fa} = 10^{-6}$, $n = 10$) (from Schleher [6], © 1976 IEEE)*

P_d	Rayleigh	Weibull	Log-normal
0·9	7·5	11·6	11·3
0·5	5·7	10·9	10·1

examined. This might be expected, since all the information contained in the input distribution is not effectively utilised.

The robustness of the median detector can be estimated by considering its performance over a range of log-normal to Weibull-to-Rayleigh clutter. This comparison is given in Table 3.13 for integration of 10 pulses with $P_{fa} = 10^{-6}$ and P_d of 0·5 and 0·9. Examination of Table 3.13 shows that the performance is reasonably robust for high probabilities of detection when a small number of samples (10) is considered.

3.4.5 Chernoff bound of optimum performance

It is of interest to compare the performance of the receivers described in the previous Sections with the optimum performance possible in Weibull clutter.

In order to make this comparison it is necessary to determine the optimum performance. This can be accomplished using a method derived from the Chernoff bounding technique.

The Chernoff bound provides an upper bound in the form of an exponential relationship. Van Tree [17] tightened the upper bound by finding a multiplication factor for the exponential relation using a central-limit-theory argument. The extension of this technique consists essentially of the development of higher-order terms in a series expansion for the performance, rather than a determination of a multiplication factor in an upper bound.

The moment generating function for the clutter hypothesis is given by:

$$\theta_{l|H_0}(s) = E(e^{sl}|H_0) = \int_{-\infty}^{\infty} e^{sl} f_l(l|H_0)\, dl \tag{3.85}$$

The semi-invariant of the likelihood ratio for n independent samples of the clutter is defined as follows:

$$\mu(s) = \ln \theta_{l|H_0}(s) = \ln \int_{-\infty}^{\infty} e^{sl} f_l(l|H_0)\, dl \tag{3.86}$$

However, the semi-invariant $\mu(s)$ cannot be found since it is expressed in terms of the unknown density function $f(l|H_0)$. But $l(v)$ is just a function of v and hence eqn. 3.86 can be written as

$$\mu(s) = n \ln \int_{-\infty}^{\infty} [f_v(v|H_1)]^s [f_v(v|H_0)]^{1-s}\, dv \tag{3.87}$$

where n is the number of independent samples, $f_v(v|\cdot)$ is the probability-density function under the hypothesis H_0 or H_1, and s a variable between 0 and 1. It was further shown [27] that the probability of false alarm for the optimum receiver can be expressed in terms of $\mu(s)$ and its derivatives as

$$P_{fa} = 1/2 \operatorname{erfc}[\alpha(s)] \exp[\alpha^2(s) + B(s)] - [\dddot{\mu}(s)/6[\ddot{\mu}(s)]^{3/2}] \\ \times \exp[B(s)]\{1/2 \operatorname{erfc} \alpha(s)[s\sqrt{\mu(s)}]^3 \exp[\alpha^2(s)] + (1/\sqrt{2\pi})[1 - s^2\ddot{\mu}(s)]\} \tag{3.88}$$

where

$$\alpha(s) = s[\ddot{\mu}(s)/2]^{1/2} \tag{3.89}$$

$$B(s) = \mu(s) - s\dot{\mu}(s) \tag{3.90}$$

An expression for P_m ($P_m = 1 - P_d$) can be found by substituting $s - 1$ for s in eqn. 3.88 through eqn. 3.90.

The procedure for finding P_{fa} and P_d is given in the following discussion. Eqn. 3.87 is evaluated by Gaussian quadrature for a particular signal value over a range of s from 0 to 1. Derivatives of $\mu(s)$ are evaluated using a spline-function numerical differentiation technique. A receiver operating curve is generated by plotting P_d versus P_{fa} for a particular signal, and Marcum–Swerling curves for the optimum receiver are generated from the receiver operating curves. Curves are presented in Fig. 3.25 that allow a comparison between practical and optimum performance in Weibull clutter, and that also allow a comparison of the optimum performance in Rayleigh, Weibull and log-normal [27] clutter.

These curves show the performance ($P_{fa} = 10^{-6}$) of the linear receiver, logarithmic receiver, binary integrator and median detector in Weibull clutter ($c = 1{\cdot}2$) for $n = 3$, 10 and 30 (independent samples), respectively. Also shown in these curves is the optimum performance in Weibull clutter ($c = 1{\cdot}2$) and Rayleigh clutter.

The performance of the binary integrator is the best of the receivers considered, approaching the optimum bound, while the performance of the median detector is the poorest in Weibull clutter. The performance of the logarithmic receiver is almost as good as the binary integrator.

Both the binary integrator and logarithmic receiver were also identified as being good receivers in log-normal clutter [27] and represent practical nonlinear receivers that provide good performance in clutter distributions that have long tails. Since both these receivers are almost optimum in Rayleigh clutter, they represent receivers that perform well over a wide range of clutter distributions.

It is of interest to compare the integration gains available from an optimum detector in Rayleigh, Weibull ($c = 1{\cdot}2$), and log-normal ($\sigma_v = 0{\cdot}693$) clutter. Table 3.14 shows the integration gain for $P_d = 0{\cdot}9$ and $P_{fa} = 10^{-6}$.

It is seen that for Weibull and log-normal clutter the situation exists whereby the integration gain is greater for noncoherent integration than it would be for coherent integration of a Gaussian distribution ($10 \log n$).

But all these results are based on the assumption that all the samples, including clutter echoes and target signals, are independent, as we mentioned before. This assumption is not true, especially for the clutter signal, and this is the difference between clutter and noise. When the samples of the same range bin but successive sweeps are integrated, not only the target signal, but also the clutter, will be integrated. This is due to the correlation between adjacent clutter echoes within the resolution cell.

Conte *et al.* [29] and Zhu [30] studied the asymptotic relative efficiency of a locally optimum detector in log-normal and Weibull clutter.

Locally optimum detectors (LOD) [31, 32], which maximise the slope of the power function at zero signal level while keeping a fixed false-alarm rate, approach optimum performance as the signal becomes vanishingly small. Satisfactory system performance, however, requires a large sample size N; but, as N increases, the LODs become more and more suboptimum [33]. Therefore, it seems more profitable to consider asymptotically optimum detectors (AOD) [33, 34], whose operating characteristics approach the optimum as N increases and the signal level vanishes. The synthesis of such detectors is based on the asymptotic expansion of likelihood ratio (LR) for sample size N approaching ∞ and signal amplitude decreasing as $1/\sqrt{N}$.

If one assumes that multiple observations are available at the receiver input in the form of signal samples plus noise samples, independent of one another and of the signal, the radar detection problem can be represented by the

Table 3.14 *Integration gain in dB for Rayleigh, Weibull and log-normal clutter (from Schleher [6], © 1976 IEEE)*

Type of clutter	n independent samples 3	10	30
Rayleigh	4·0	8·0	11·2
Weibull ($c = 1{\cdot}2$)	6·2	12·3	17·6
Log-normal ($\sigma_v = 0{\cdot}693$)	11·8	18·2	22·6

following hypothesis test:

$$H_0: \tilde{R} = \tilde{N}$$

$$H_1: \tilde{R} = a\tilde{S} + \tilde{N} \tag{3.91}$$

where $\tilde{R}^T = (\tilde{r}_1, \tilde{r}_2, \ldots, \tilde{r}_N)$, $\tilde{N}^T = (\tilde{n}_1, \tilde{n}_2, \ldots, \tilde{n}_N)$ and $a\tilde{S}^T = a(\tilde{s}_1, \tilde{s}_2, \ldots, \tilde{s}_N)$ are the complex envelope sample vectors of the received narrowband waveform, noise and signal, respectively; $\tilde{r}_i = z_i e^{j\varphi_i}$, $i = 1, 2, \ldots, N$; z_i and φ_i are the ith sample of the received waveform envelope and phase, respectively and a represents the average voltage (signal-to-noise ratio, SNR); $\tilde{s}_i = s_i e^{j\theta_i}$, the ith sample of the signal complex envelope normalised with respect to a and s_i and θ_i are the ith sample of the envelope and phase respectively. It is assumed that the noise envelope and phase are statistically independent; the noise phase is uniformly distributed in $[0, 2\pi]$ and the noise samples $\{\tilde{n}_i, i = 1, 2, \ldots, N\}$ are independent and identically distributed with an envelope PDF expressed as $p_o(\cdot)$. It should be noted that the assumption of independent samples is not true in many practical cases—only if decorrelation techniques, such as frequency agility, are employed.

When H_0 is true, the joint PDF of z_i and φ_i is

$$f_o(z_i, \varphi_i) = \frac{1}{2\pi} p_o(z_i), \quad z_i \geqslant 0, \quad 0 \leqslant \varphi_i < 2\pi \tag{3.92}$$

Considering the real and imaginary parts of $\tilde{r}_i$ and transforming the variable, we obtain the conditional joint PDF of z_i and φ_i under H_1 as

$$f_1(z_i, \varphi_i | \theta_2) = \frac{z_i p_o(\sqrt{z_i^2 - 2as_i z_i \cos(\varphi_i - \theta_i) + a^2 s_i^2})}{2\pi\sqrt{z_i^2 - 2as_i z_i \cos(\varphi_i - \theta_i) + a^2 s_i^2}},$$

$$z_i \geqslant 0, \quad 0 \leqslant \varphi_i < 2\pi \tag{3.93}$$

Three cases of reception are considered, namely, coherent pulse train signals with known initial phase (s_i and θ_i are known), coherent pulse train signals with unknown initial phase (s_i is known), and incoherent pulse train signals, (s_i is known and θ_i's are independent and uniformly distributed in $[0, 2\pi]$).

(i) In the case of coherent pulse train signals with known initial phase, s_i and θ_i are known. The likelihood ratio is

$$\Lambda(\tilde{R}) = \prod_{i=1}^{N} \frac{f_i(z_i, \varphi_i | \theta_i)}{f_o(z_i, \varphi_i)} \tag{3.94}$$

This is a case of coherent reception. The locally optimum detector (LOD) structure is determined by the test of the likelihood ratio under two hypothesis H_0 and H_1 according the Neyman–Pearson criterion; i.e. the required SNR is a minimum for two given kinds of error probability. This can be written as

$$\left.\frac{\partial \Lambda(\tilde{R})}{\partial a}\right|_{a=0} \underset{H_0}{\overset{H_1}{\gtrless}} T \tag{3.95}$$

where T is a decision threshold, which is determined by the given false-alarm probability. From eqn. 3.93, we obtain

$$\left.\frac{\partial f_1(z_i, \varphi_i|\theta_i)}{\partial a}\right|_{a=0} = \frac{s_i \cos(\varphi_i - \theta_i)}{2\pi}\left[\frac{p_o(z_i)}{z_i} - p'_o(z_i)\right] \tag{3.96}$$

Using eqn. 3.96, we obtain from eqn. 3.94

$$\left.\frac{\partial \Lambda(\tilde{R})}{\partial a}\right|_{a=0} = \mathrm{Re}\left\{\sum_{i=1}^{N} g_{c,l_o}(z_i)e^{j\varphi_i}\tilde{s}_i^*\right\} \tag{3.97}$$

where

$$g_{c,l_o}(z) = \frac{1}{z} - \frac{p'_o(z)}{p_o(z)} \tag{3.98}$$

Eqn. 3.97 shows a structure of narrowband zero-memory nonlinearity (ZNL) correlation, taking the real part. The LOZNL is defined by eqn. 3.98.

(ii) In the case of coherent-pulse-train signals with unknown initial phase, s_i is known, and $\theta_i = \theta_{io} + \Delta\theta$, where θ_{io} is known and $\Delta\theta$ is uniformly distributed in $[0, 2\pi]$. Then the likelihood ratio is

$$\Lambda(\tilde{R}) = \frac{\int_0^{2\pi}\left[\prod_{i=1}^{M} f_1(z_i, \varphi_1|\theta_i)\right] d(\Delta\theta)/2\pi}{\prod_{i=1}^{M} f_o(z_i, \varphi_i)} \tag{3.99}$$

This is a case of incoherent reception; the LO detector structure is given by

$$\left.\frac{\partial^2 \Lambda(\tilde{R})}{\partial a^2}\right|_{a=0} \underset{H_0}{\overset{H_1}{\gtrless}} T \tag{3.100}$$

It can be shown that

$$\left.\frac{\partial^2 \Lambda(\tilde{R})}{\partial a^2}\right|_{a=0} = \frac{1}{2}\left\{\left|\sum_{i=1}^{M} g_{c,l_o}(z_i)s_i e^{j(\varphi_i - \theta_{io})}\right|^2 + \sum_{i=1}^{M} g_{ic,l_o}(z_i)s_i^2\right\} \tag{3.101}$$

where $g_{c,l_o}(\cdot)$ is the same as that given in eqn. 3.98, and $g_{ic,l_o}(\cdot)$ is defined by

$$g_{ic,l_o}(z) = \frac{p''_o(z)}{p_o(z)} + \frac{p'_o(z)}{zp_o(z)} - \left[\frac{p'_o(z)}{p_o(z)}\right]^2 \tag{3.102}$$

The first term on the right hand side of eqn. 3.101 represents a coherent branch with a structure of narrowband ZNL-correlation-envelope detection, the LOZNL of which is the same as that in case of coherent pulse-train signals with known initial phase, and is given by eqn. 3.98. The second term shows an incoherent branch with a structure of envelope detection-baseband ZNL correlation, and its LOZNL is defined by eqn. 3.102. As the incoherent branch discards phase information, the coherent branch plays a major role.

(iii) In the case of incoherent pulse train signals, s_i is known and θ_i's are independent and uniformly distributed in $[0, 2\pi]$. The likelihood ratio is

$$\Lambda(R) = \frac{\int_{\{\Theta\}} \left[\prod_{i=1}^{N} f_1(z_i, \varphi_i|\theta_i)\right] d\Theta/(2\pi)^N}{\prod_{i=1}^{N} f_o(z_i, \varphi_i)} \tag{3.103}$$

where $\Theta^T = (\theta_1, \theta_2, \ldots, \theta_N)$ is the sample vector of signal phase. This is an incoherent reception case, and the LO detector structure is determined by eqn. 3.100. Using eqns. 3.92 and 3.103, it can be shown that

$$\left.\frac{\partial^2 \Lambda(\tilde{R})}{\partial a^2}\right|_{a=0} = \frac{1}{2}\sum_{i=1}^{N} g_{i,l_o}(z_i)s_i^2 \tag{3.104}$$

where

$$g_{i,l_o}(z) = \frac{p_o''(z)}{p_o(z)} - \frac{p_o'(z)}{zp_o(z)} + \frac{1}{z^2} \tag{3.105}$$

Eqn. 3.104 represents a structure of envelope detection-baseband ZNL correlation, and its LOZNL is given by eqn. 3.105.

It can be seen from the above that all the LO detector structures for the three classes of signals (in the case of coherent pulse-train signals with unknown initial phase, it means the coherent branch is considered) corresponds to incorporating proper LOZNL processing into the Neyman–Pearson optimum detector in narrowband Gaussian noise for the same class of signals.

The structures of these three kinds LO detectors are shown in Fig. 3.26. All these LO detectors have the same class of structure but different ZNL characteristic $g(\cdot)$ and filter weighting sequence $\{h_i\}$. In the general LO detector structures of Fig. 3.26 the outputs of M independent diversity channels with the same signal are summed.

The detection performances of these detectors are measured with asymptotic relative efficiency (ARE) under large-sample-size and small-signal conditions. The ARE of a detector A with respect to detector B can be expressed as the ratio of their respective efficacies e_A and e_B, i.e.

$$\text{ARE}_{A,B} = \frac{e_A}{e_B} \tag{3.106}$$

where e_A and e_B are the efficacy of detector A and detector B, respectively. The efficacy of a detector is defined as

$$e(t) = \lim_{N\to\infty} \frac{\left[\left.\frac{\partial}{\partial s} E(t/s)\right|_{s=0}\right]^2}{N\sigma_o(t)} \tag{3.107}$$

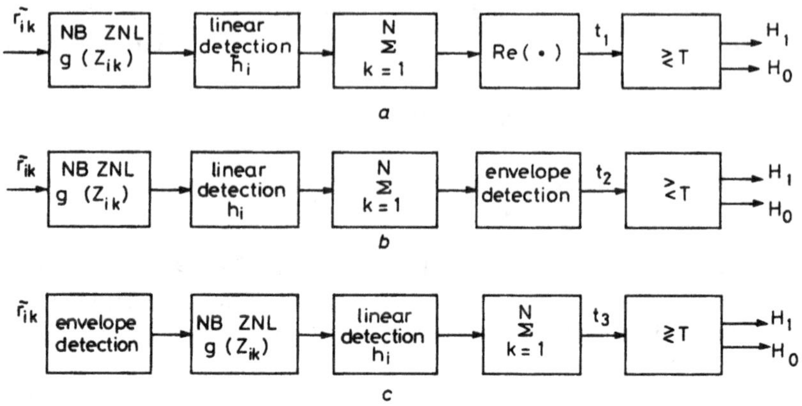

Fig. 3.26 ***General detector structure of three types of signals***
(*a*) LO detector for coherent pulse-train signals with known initial phase
(*b*) LO detector for coherent pulse-train signals with unknown initial phase
(*c*) LO detector for incoherent pulse-train signals.

where t is the test statistic, s is the input signal-to-noise ratio, $\sigma_o(t)$ is the variance of t under no-signal condition and $E(t/s)$ is the mean of t under the signal-to-noise ratio s.

The test statistic for the detector of Fig. 3.26*a* is

$$t_1 = \mathrm{Re}\left\{\sum_{k=1}^{M}\sum_{i=1}^{N} g(z_{ik})e^{i\varphi_{ik}}\tilde{h}_{N+1-i}\right\} \tag{3.108}$$

which is asymptotically normal. Using eqns. 3.92 and 3.96, we find the efficacy of t_1 is

$$e_1 = \lim_{M\to\infty} \frac{\left[\left.\dfrac{\partial E(t_1|a)}{\partial a}\right|_{a=0}\right]^2}{M\,\mathrm{var}_0(t_1)}$$

$$= \frac{\left\{\displaystyle\int_0^\infty g(z)\left[\frac{p_o(z)}{z} - p_o'(z)\right]dz\right\}^2}{2\displaystyle\int_0^\infty g^2(z)p_o(z)\,dz} \; \frac{\left\{\displaystyle\sum_{i=1}^{N} \mathrm{Re}(\tilde{s}_i\tilde{h}_{N+1-i})\right\}^2}{\displaystyle\sum_{i=1}^{N} h_i^2} \tag{3.109}$$

where $E(t_1|a)$ denotes the conditional expectation of t_1 conditioned upon the SNR a and $\mathrm{var}_0(t_1)$ is the variance of t_1 when $a = 0$.

The test statistic for the detector of Fig. 3.26*b* is

$$t_2 = \left|\sum_{k=1}^{M}\sum_{i=1}^{N} g(z_{ik})e^{j\varphi_{ik}}\tilde{h}_{N+1-i}\right| \tag{3.110}$$

which is not asymptotically normal, and thus the common concept of efficacy and ARE cannot be used directly.

It can be shown that, when H_0 is true and N is large, t_2 is of an approximate Rayleigh distribution with the following PDF:

$$p(t_2|H_0) = \frac{t_2}{\sigma^2} \exp\left(-\frac{t_2^2}{2\sigma^2}\right), \quad t_2 > 0 \tag{3.111}$$

where

$$\sigma^2 = M \int_0^\infty g^2(z) p_o(z)\, dz \sum_{i=1}^{N} \frac{h_i^2}{2} \tag{3.112}$$

When H_1 is true with large M and small a, t_2 is of an approximate Rice distribution with the PDF as follows:

$$p(t_2|H_1) = \frac{t_2}{\sigma^2} \exp\left[-\frac{t_2^2 + \xi^2}{2\sigma^2}\right] I_o\left(\frac{\xi t_2}{\sigma^2}\right), \quad t_2 \geqslant 0 \tag{3.113}$$

where

$$\xi = \mathrm{Ma} \int_0^\infty g(z)\left[\frac{p_o(z)}{z} - p_o'(z)\right] dz \left|\sum_{i=1}^{N} \frac{\tilde{s}_i \tilde{h}_{N+1-i}}{2}\right]^2 \tag{3.114}$$

For the fixed probabilities of false alarm and detection, ξ^2/σ^2 must remain unchanged when M increases and a decreases. From eqns. 3.112 and 3.114, we have

$$\xi^2/\sigma^2 = e_2 \mathrm{Ma} \tag{3.115}$$

where

$$e_2 = \frac{\left\{\int_0^\infty g(z)\left[\frac{p(z)}{z} - p_o'(z)\right] dz\right\}^2}{2\int_0^\infty g^2(z) p_o(z)\, dz} \frac{\left|\sum_{i=1}^{N} \tilde{s}_i \tilde{h}_{N+1-i}\right|}{\sum_{i=1}^{N} h_i^2} \tag{3.116}$$

By an agreement similar to that for theorems 1 and 2 of [35], we can take e_2 of eqn. 3.116 as the efficacy of t_2, and also express the ARE of two detectors as the ratio of their efficacies.

The test statistic for the detector of Fig. 3.26*c* is

$$t_3 = \sum_{k=1}^{M} \sum_{i=1}^{N} g(z_{ik}) H_{N+1-i} \tag{3.117}$$

which is asymptotically normal. Using eqns. 3.92 and 3.93, it can be shown that the efficacy of t is

$$e_3 = \lim_{M\to\infty} \frac{[\partial E(t_3|a)/\partial a|_{a=0}]^2}{M\,\mathrm{var}_0(t_3)}$$

$$= \frac{\left\{\int_0^\infty g(z)\left[p_o''(z) - \frac{p_o'(z)}{z} + \frac{p_o(z)}{z^2}\right] dz\right\}^2}{4\left\{\int_0^\infty g^2(z)p_o(z)\,dz - \left[\int_0^\infty g(z)p_o(z)\,dz\right]^2\right\}} \frac{\left(\sum_{i=1}^N s_i^2 h_{N+1-i}\right)^2}{\sum_{i=1}^N h_i^2} \qquad (3.118)$$

As can be seen from eqns. 3.109, 3.116 and 3.118, the efficacies of the three classes of detectors can all be expressed as a product of two terms. The first term depends only on ZNL and the noise-envelope statistics. The second term depends only on the filter and the signal, and is a maximum when the filter matches the signal (in the incoherent pulse-train-signal case, the filter matches the square of the signal). Furthermore, e_1 and e_2 have the same first term, which is

$$e_c = \frac{\left\{\int_0^\infty g(z)\left[\frac{p(z)}{z} - p_o'(z)\right] dz\right\}^2}{2\int_0^\infty g^2(z)p_o(z)\,dz} \qquad (3.119)$$

The first term of e_3 is

$$e_i = \frac{\left\{\int_0^\infty g(z)\left[p_o''(z) - \frac{p_o'(z)}{z} + \frac{p_o(z)}{z^2}\right] dz\right\}^2}{4\left\{\int_0^\infty g^2(z)p_o(z)\,dz - \left[\int_0^\infty g(z)p_o(z)\,dz\right]^2\right\}} \qquad (3.120)$$

Thus e_c and e_i can be seen as the efficacies of the detectors for detecting coherent and incoherent pulse train signals, respectively.

Now, according to eqn. 3.119, the expressions for the efficacies of several detectors for detecting coherent pulse train signals can be derived.

(i) *For the LO detector*: $g(z) = g_{c,l_o}(z)$ is given by (3.98); the efficacy is

$$e_{c,lo} = \frac{1}{2}\int_0^\infty g_{c,lo}^2(z)p_o(z)\,dz \tag{3.121}$$

(ii) *For the linear detector*: $g(z) = z, z \geqslant 0$; the efficacy is

$$e_{c,ln} = \frac{\left\{\int_0^\infty [p_o(z) - zp_o'(z)]\,dz\right\}^2}{2E_0(z^2)} \tag{3.122}$$

where $E_0(z^2)$ represents the mean square value of noise envelope.

If $zp_o(z)|_o^\infty = 0$, we have

$$e_{c,ln} = \frac{2}{E_0(z^2)} \tag{3.123}$$

(iii) *For the Dicke-fix detector*: $g(z) = 1, z \geqslant 0$; the efficacy is

$$e_{c,Df} = \frac{\left\{\int_0^\infty \left[\frac{p_o(z)}{z} - p_o'(z)\right] dz\right\}^2}{2} \tag{3.124}$$

When $p_o(z)|_0^\infty = 0$, we obtain

$$e_{c,Df} = \frac{\left\{\int_0^\infty \frac{p_o(z)}{z}\,dz\right\}^2}{2} \tag{3.125}$$

(iv) *For the logarithmic detector*: $g(z) = \ln(bz), z \geqslant 0$, where b is a constant greater than zero; the efficacy is

$$e_{c,lg} = \frac{\left\{\int_0^\infty \ln(bz)\left[\frac{p_o(z)}{z} - p_o'(z)\right] dz\right\}^2}{2\int_0^\infty [\ln(bz)]^2 p_o(z)\,dz} \tag{3.126}$$

Similarly, expressions for the efficacies of several detectors for detecting incoherent pulse train signals can be derived on the basis of eqn. 3.120.

(*a*) *For the LO detector*: $g(z) = g_{i,lo}(z)$ is given by eqn. 3.105; the efficacy is

$$e_{i,lo} = \frac{1}{4}\int_0^\infty g_{i,lo}^2(z)p_o(z)\,dz \tag{3.127}$$

(*b*) *For the square-law detector*: $g(z) = z^2, z \geqslant 0$; the efficacy is

$$e_{i,sq} = \frac{\left\{\int_0^\infty [z^2p_o''(z) - zp_o'(z) + p_o(z)]\,dz\right\}^2}{4\,\mathrm{var}_0(z^2)} \tag{3.128}$$

If $z^2p_o(z)|_0^\infty = 0$ and $zp_o(z)|_0^\infty = 0$, we have

$$e_{i,sq} = \frac{4}{\mathrm{var}_0(z^2)} \tag{3.129}$$

(*c*) *For the linear detector*: $g(z) = z, z \geqslant 0$; the efficacy is

$$e_{i,ln} = \frac{\left\{\int_0^\infty \left[zp_o''(z) - p_o'(z) + \frac{p_o(z)}{z}\right] dz\right\}^2}{4\,\mathrm{var}_0(z)} \tag{1.130}$$

Since $\lim_{z\to\infty} p_o(z) = 0$ is always true, thus, as long as $p_o(z)|_{z=0}$ is well defined, eqn. 3.130 can be reduced to

$$e_{i,ln} = \frac{\left\{\int_0^\infty \frac{p_o(z)}{z}\,dz + 2p_o(z)\Big|_{z=0}\right\}^2}{4\,\mathrm{var}_0(z)} \tag{3.131}$$

(*d*) *For the logarithmic detector*: $g(z) = \ln(bz), z \geqslant 0$, where b is a constant greater than zero; the efficacy is

$$e_{i,lg} = \frac{\left\{\int_0^\infty \ln z\left[p_o''(z) - \frac{p_o'(z)}{z} + \frac{p(z)}{z^2}\right] dz\right\}^2}{4\left\{\int_0^\infty (\ln z)^2 p_o(z)\,dz - \left[\int_0^\infty \ln z p_o(z)\,dz\right]^2\right\}} \tag{3.132}$$

(*e*) *For the binary integration detector*: $g(z) = u(z - T_1), z > 0$, where $u(\cdot)$ denotes the unit step function and T_1 the first threshold; the efficacy is

$$e_{i,bn} = \frac{\left[\frac{p_o(T_1)}{T_1} - p_o'(T_1)\right]^2}{4[F_o(T_1) - F_o^2(T_1)]} \tag{3.133}$$

where $F_o(T_1) = \int_0^{T_1} p_o(z)\,dz$. In deriving eqn. 3.133, it is assumed that $\lim_{z\to\infty} p_o(z) = 0$ and $\lim_{z\to\infty} p'_o(z)/z = 0$.

(vi) *For the median detector*: It can be considered as a special case of the binary integration detector. Setting $T = z_m$, where z_m is the median of the noise envelope, from (3.133) the efficacy of the median detector can be expressed as

$$e_{i,md} = \left[\frac{p_o(z_m)}{z_m} - p'_o(z_m)\right]^2 \tag{3.134}$$

Since $p_o(z)$ is the PDF of the noise envelope, the efficacies of all these detectors depend on the amplitude distribution of the clutter. This means that these detectors have different efficacies in different clutters.

Now let us consider, in particular, the detection in Weibull clutter. The PDF of Weibull clutter is

$$p_o(z) = \frac{\alpha}{\beta}\left(\frac{z}{\beta}\right)^{\alpha-1} \exp\left[-\left(\frac{z}{\beta}\right)^{\alpha}\right], \quad z \geqslant 0 \tag{3.135}$$

where α is the shape parameter and β the scale parameter. Both parameters are positive. The smaller the α is, the more slowly the tail of PDF curve decays. When $\alpha = 2$, eqn. 3.135 reduces to Rayleigh distribution.

Using the results obtained above, the expressions for the LOZNLs can be developed and the formulas of the detector efficacies for detecting coherent and incoherent pulse-train signals in Weibull clutter can be derived as summarised in Table 3.15.

The numerical results for the LOZNLs and AREs are given in Figs. 3.27—3.31.

The AREs are evaluated with respect to the linear (square-law) detector in the case of detecting coherent (incoherent) pulse-train signals. It is well known that the two detectors are, respectively, the Neyman–Pearson optimum

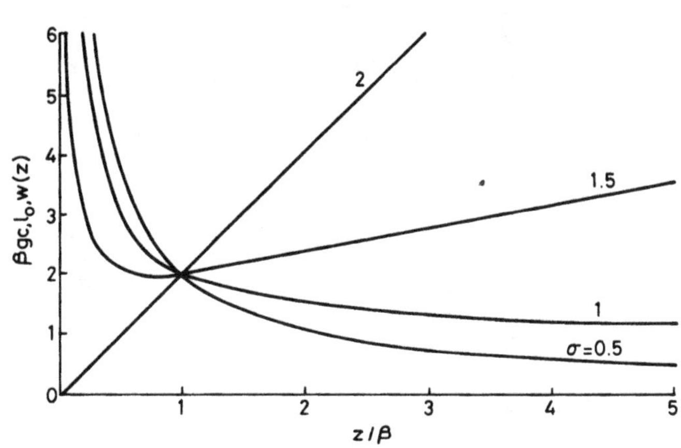

Fig. 3.27 *Normalised locally optimum nonlinearity for detecting coherent pulse-train signals in Weibull clutter*

Table 3.15 *LOZNL and detection efficacies of different detectors in Weibull clutter*

	Signal	Coherent pulse train	Incoherent pulse train
	LOZNL	$\frac{2-\alpha}{z}+\frac{\alpha}{\beta}\left(\frac{z}{\beta}\right)^{\alpha-1}$	$\frac{(\alpha-2)^2}{z^2}-\frac{3\alpha^2-4\alpha}{\beta^2}\left(\frac{z}{\beta}\right)^{\alpha-2}+\frac{\alpha^2}{\beta^2}\left(\frac{z}{\beta}\right)^{2\alpha-2}$
Detector efficacy	LO	$\begin{cases}\beta^{-2}\frac{\alpha(\alpha-2)}{2}\Gamma\left(1-\frac{2}{\alpha}\right), & \alpha>2\\ 2\beta^{-2}, & \alpha=2\end{cases}$	$\begin{cases}\beta^{-4}\frac{(\alpha-2)(5\alpha^3-26\alpha^2+28\alpha-8)}{4}\Gamma\left(1-\frac{4}{\alpha}\right), & \alpha>4\\ 4\beta^{-4}, & \alpha=2\end{cases}$
	Linear	$\beta^{-2}\frac{2}{\Gamma\left(1+\frac{2}{\alpha}\right)}$	$\beta^{-4}\frac{\Gamma^2\left(1-\frac{1}{\alpha}\right)}{4\left[\Gamma\left(1+\frac{2}{\alpha}\right)-\Gamma^2\left(1+\frac{1}{\alpha}\right)\right]}, \quad \alpha>1$
	Square law		$\beta^{-4}\frac{4}{\Gamma\left(1+\frac{4}{\alpha}\right)-\Gamma^2\left(1+\frac{2}{\alpha}\right)}$

Dicke fix	$\beta^{-2}\dfrac{\Gamma^2\left(1-\dfrac{1}{\alpha}\right)}{2}, \quad \alpha>1$	
Logarithmic*	$\beta^{-2}\Gamma^2\left(1-\dfrac{1}{\alpha}\right)\left\{\dfrac{1}{2}+\dfrac{3}{\pi^2}\left[\psi\left(1-\dfrac{1}{\alpha}\right)+\alpha+C\right]^2\right\}, \quad \alpha>1$	$\begin{cases} 0, & \alpha>2 \\ \dfrac{24}{\pi^2}\beta^{-4}, & \alpha=2 \end{cases}$
Binary integration		$\beta^{-4}\dfrac{\left\{\alpha\left(\dfrac{T_1}{\beta}\right)^{\alpha-2}\left[2-\alpha+\alpha\left(\dfrac{T_1}{\beta}\right)^{\alpha}\right]\right\}^2}{4\left\{\exp\left[\left(\dfrac{T_1}{\beta}\right)^{\alpha}\right]-1\right\}}$
Median		$\beta^{-4}\dfrac{\{\alpha(\ln 2)^{1-2/\alpha}[2-\alpha+\alpha\ln 2]\}^2}{4}$

*$\psi(\cdot)$—ψ-function, C—Euler constant

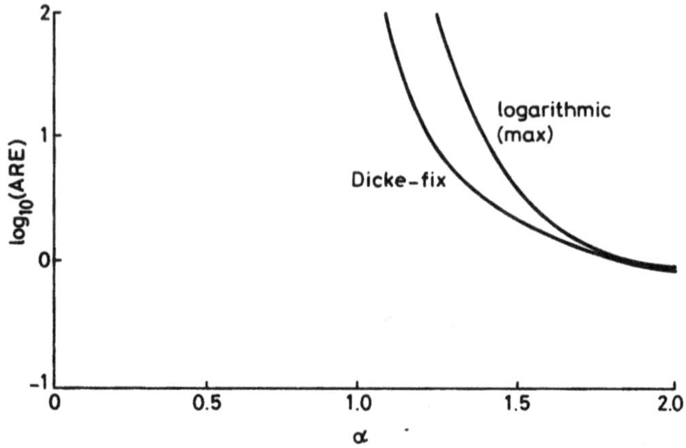

Fig. 3.28 *ARE of detectors with respect to linear detector, for detecting coherent pulse-train signals in Weibull clutter*

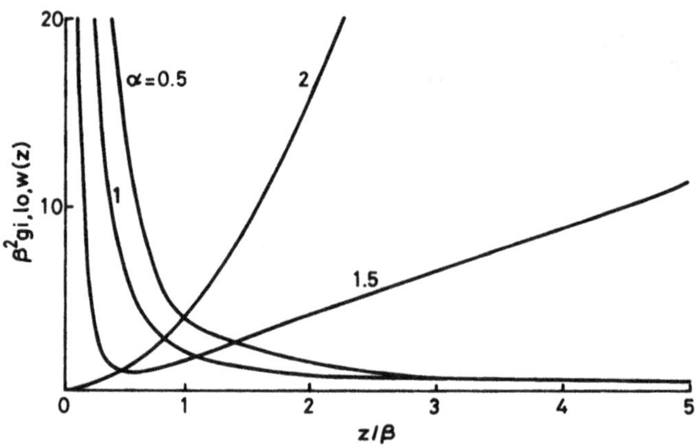

Fig. 3.29 *Normalised locally optimum nonlinearity for detecting incoherent pulse-train signals in Weibull clutter*

detectors for detecting coherent and incoherent pulse-train signals in narrow-band Gaussian noise.

The Figures show that the form of LOZNLs depends on the clutter PDF in critical ways. The *a priori* knowledge about the exact clutter PDF is often lacking. So it is difficult to implement LO detectors. Fortunately, many practical detectors, such as the Dicke-fix, logarithmic, binary integration and median detectors, whose ZNL can suppress more effectively the tail of the clutter envelope distribution, are apparently asymptotically efficient over a broad range of the clutter-envelope density parameter α and σ. Their ARE relative to square-law detectors is in many cases, especially for small α, greater than 1.

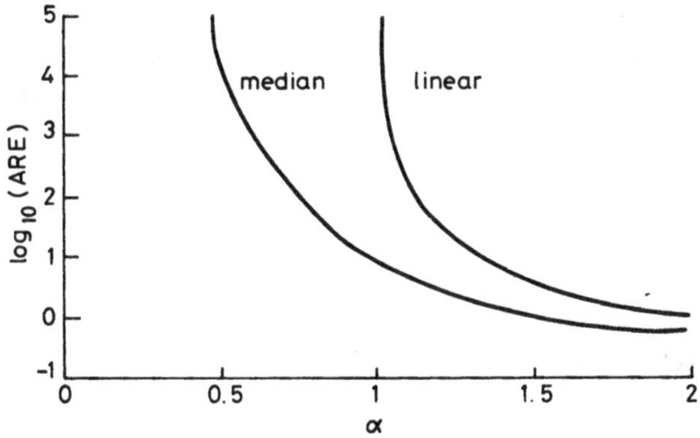

Fig. 3.30 *ARE of detectors with respect to square-law detector, for detecting incoherent pulse-train signals in Weibull clutter*

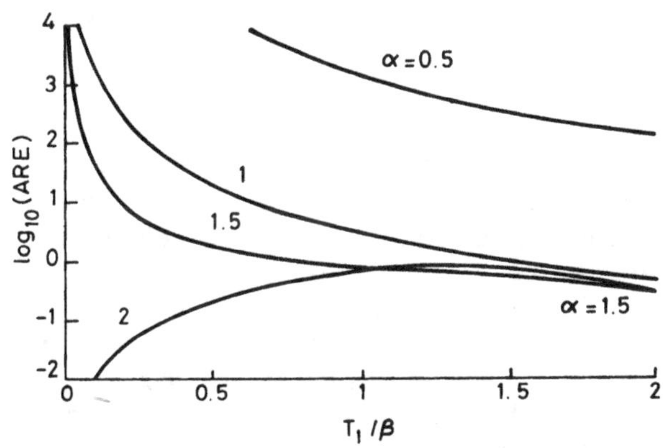

Fig. 3.31 *ARE of binary integration detector with respect to square-law detector, for detecting incoherent pulse-train signals in Weibull clutter*

3.5 Detection performance of CFAR detector in Weibull clutter

The difference between Weibull clutter and Rayleigh clutter is the long tail in the case of the former density function. This will cause false alarms for fixed threshold. Therefore, the adaptive CFAR detector is desirable, as we mentioned before. However, the CFAR processes will introduce CFAR detection loss owing to the threshold being adjusted to the mean level of the clutter background. Since the mean of the clutter is estimated from a limited number of samples along the range direction, the estimated value fluctuates with time.

Therefore, it is necessary to study the CFAR detection loss for different CFAR detectors in Weibull clutter.

3.5.1 CFAR loss of log t test

The log t detector proposed by Goldstein is widely used in marine radar [36] to provide adaptive CFAR processing in an extremely non-stationary environment. We have discussed the CFAR properties and the threshold characteristic in Section 3.2. Now we will discuss the CFAR loss of log t test.

If the signal-to-clutter power ratio γ is defined as

$$\gamma = E(A_s^2)/E(A_c^2)$$

where A is the amplitude of the signal, and A_c is the amplitude of clutter, we find that, for Weibull clutter,

$$\gamma = \frac{2\sigma_s^2}{\{v^2\Gamma[1+(2/\eta)]\}} \tag{3.136}$$

where σ_s^2 is the average signal power, v is the scale parameter, and η is the shape parameter of Weibull distribution as defined in eqn. 3.4. When the approximation for large γ is used, the asymptotic detection probability in Weibull clutter can be expressed in terms of γ_∞ as follows:

$$\ln P_d = \frac{-\exp[(2/\eta)(1{\cdot}284T_\infty + \chi(1))]}{\{\gamma_\infty\Gamma[1+(2/\eta)]\}} \tag{3.137}$$

When the number of reference cells is finite, the detection probability of the log t test no longer has the simple form of eqn. 3.137. However, a reasonable approximation to the detector performance can be obtained using the same formula, except that, in place of the asymptotic threshold T_∞, we insert the actual threshold required to obtain the desired P_{fa} (Fig. 3.6). This approximation improves for any given N as the signal-to-clutter ratio increases.

No extensive analysis of this approximation has been made. However, its utility was justified on the basis of limited verification by computer simulation and the fact that it leads to simple results which are indicative of the actual performance.

Letting T_N, γ_N, T_∞ and γ_∞ be as defined before, and using the approximation discussed above, (3.137) can be applied to determine the CFAR detection loss.

The results are, for Weibull clutter,

$$\left(\frac{\gamma_N}{\gamma_\infty}\right)(\text{dB}) = \frac{11{\cdot}15(T_N - T_\infty)}{\eta} \quad (\text{dB}) \tag{1.138}$$

For example, with $\eta = 2$ (Rayleigh clutter) and $N = 50$, the CFAR loss for a design $P_{fa} = 10^{-4}$ is 2·5 dB. It is approximately the same as Hansen's results (see Fig. 3.5).

3.5.2 Detection performance of non-parametric CFAR detector

Non-parametric CFAR detectors can maintain the false alarm constant for any input density function. It is a 'distribution free' type CFAR detector.

However, it suffers from high CFAR loss. The reason is that the maximum value of rank is equal to the number of reference cells, which is equivalent to hard-limiting the input signal. Thus the CFAR loss is introduced.

Zhu [37] studied the asymptotic detection performance of the non-parametric quantised-rank CFAR detector (QRD).

It can be shown that the efficacy of QRD is

$$e_{QR} = \frac{N+1-T_1}{\sigma^4 T_1}\left(\sum_{i=N+2-T_1}^{N+1} 1/i\right)^2, \quad 1 \leqslant T_1 \leqslant N \tag{3.139}$$

where N is the number of reference cells, σ^2 is the average power of Gaussian noise, and T_1 is the quantisation threshold. The optimum-rank quantisation threshold (ORQT), which maximises the e in Gaussian noise when $N \to \infty$, is approximately [24]

$$T_{1\,opt} = 0{\cdot}8(N+1) \tag{3.140}$$

The optimum parametric detector for Gaussian noise and small signal-to-noise ratio is a square-law detector (SLD). Its efficacy in Gaussian noise

$$e_{SL} = 1/\sigma^4 \tag{3.141}$$

In the case of noncoherent detection, the asymptotic loss $L_{A,B}$ of a detector A with respect to a detector B is related to their efficacies e_A and e_B by

$$L_{A,B} = (e_A/e_B)^{-1/2} \tag{3.142}$$

or

$$L_{A,B} = (\mathrm{ARE})^{-1/2} \tag{3.143}$$

From (3.139)—(3.142) the ARE and $L_{QR,SL}$ can be calculated.

Hansen [22] also calculated the ARE and $L_{RS,SL}$ of the rank sum nonparametric detector, which he called generalised-sign (GS) test detector.

The results of ARE and $L_{A,B}$ for these two types of nonparametric detectors are listed in Table 3.16.

For radar application it is of interest to calculate the detection performance of nonparametric CFAR detectors with a small number of samples.

Hansen [22] calculated the detection loss of a rank-sum non-parametric CFAR detector with finite number of samples. The results are shown in Figs. 3.32*a* and *b* for non-fluctuating signal and pulse-to-pulse Rayleigh fluctuating (Swerling case II), respectively. The P_{fa} and P_d for both Figures are equal to 0·5 and 10^{-6}. The loss is shown as a function of the number of hits M integrated, with the number of reference cells N as parameter.

It can be seen from these Figures that the detection loss of a non-parametric CFAR detector compared to the optimum parametric procedure is larger for a small number of reference cells N and a small number of pulses M integrated. When $M \to \infty$, the detection loss is equal to an asymptotic value L_∞, which is shown in Table 3.16. For a pulse-to-pulse Rayleigh fluctuating

Table 3.16 *Asymptotic performance of nonparametric detector*

	ARE		*L*(dB)	
N	QRD	RSD(GS)	QRD	RSD(GS)
1	0·25	0·25	3·01	3·0
2	0·346	0·375	2·30	2·1
4	0·41	0·5	1·93	1·5
8	0·50	0·6	1·49	1·1
16	0·565	0·667	1·24	0·9
	0·649	0·75	0·94	0·6

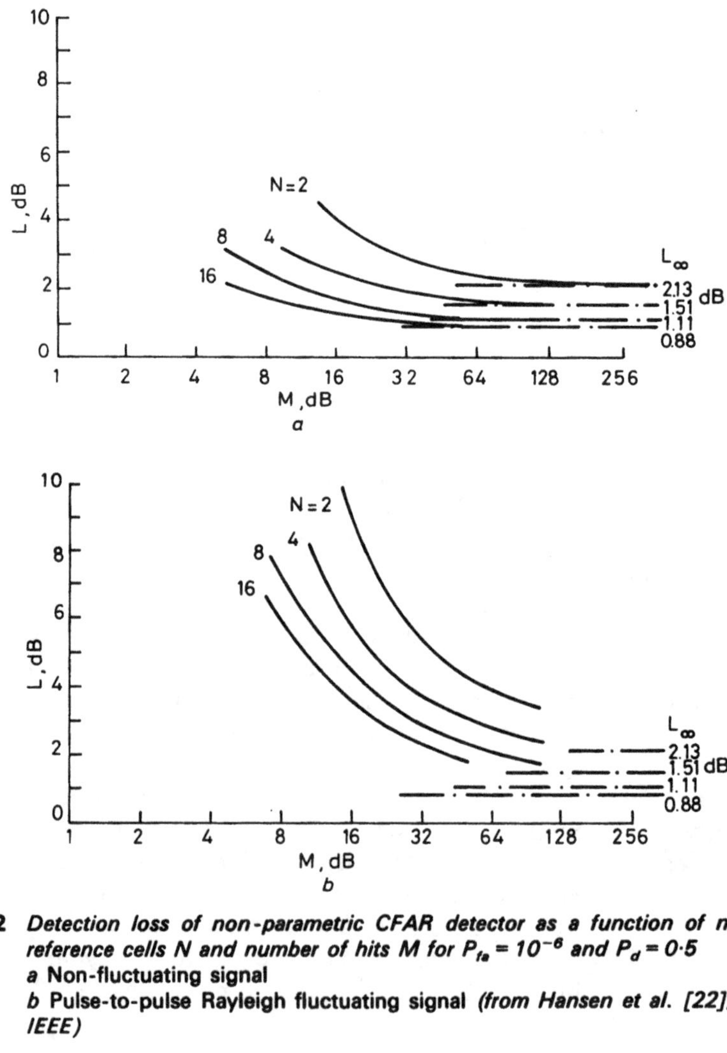

Fig. 3.32 *Detection loss of non-parametric CFAR detector as a function of number of reference cells N and number of hits M for $P_{fa} = 10^{-6}$ and $P_d = 0{\cdot}5$*
a Non-fluctuating signal
b Pulse-to-pulse Rayleigh fluctuating signal *(from Hansen et al. [22], © 1971 IEEE)*

signal the loss is significantly larger for small values of M, even though the asymptotic performance is the same in the two cases.

Unfortunately, all these values are evaluated for Gaussian noise. It is evident that, if the clutter is of Rayleigh distribution, it is not necessary to use the non-parametric CFAR detector. Since most of the clutter can be modelled as a Weibull distribution, it is of interest to compare the detection performance of a nonparametric detector with a parametric detector in a Weibull-clutter environment. For the same reason the detection performance for a small number of pulse M is desired.

Since an analytical solution cannot be obtained, most authors use Monte Carlo simulation to determine the small-sample performance of the nonparametric CFAR detector.

In the case of a quantised rank detector, Fang [38] pointed out that the detection performance of the QRD is very sensitive to the rank-quantisation threshold. Fig. 3.33 shows the effects of the rank-quantisation threshold on the detection probability when the shape parameter of the Weibull distribution is equal to 1·2, $P_{fa} = 10^{-6}$ and $M = 30$.

Zhu [37] pointed out that the optimum-rank quantisation threshold (ORQT) of the QRD, which maximises the detection probability P_d for given P_{fa} and signal-to-median clutter, is also very sensitive to the shape parameter α of the Weibull distribution. The range of ORQT is 0·8N to N for $\alpha = 2$. When α decreases, the range of ORQT extends downward; that is 0·53N to N

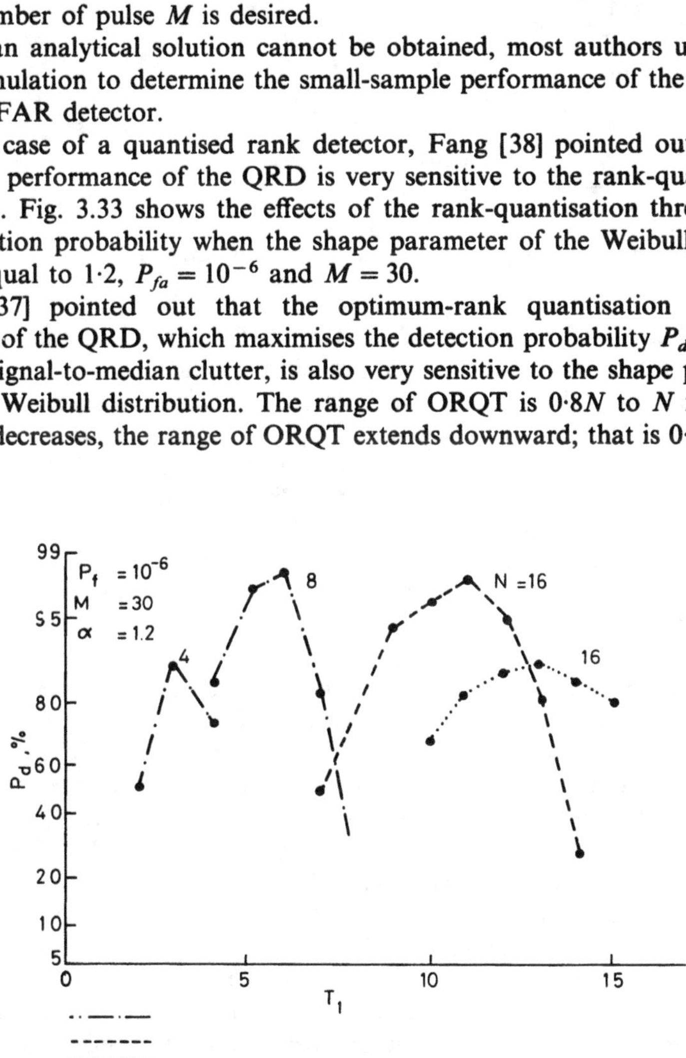

Fig. 3.33 *The effect of quantisation threshold on P_d*
—·—·— steady, $s/c = 8$ dB
---- target, $s/c = 7$ dB
······ log-normal target, $s/c = 12$ dB

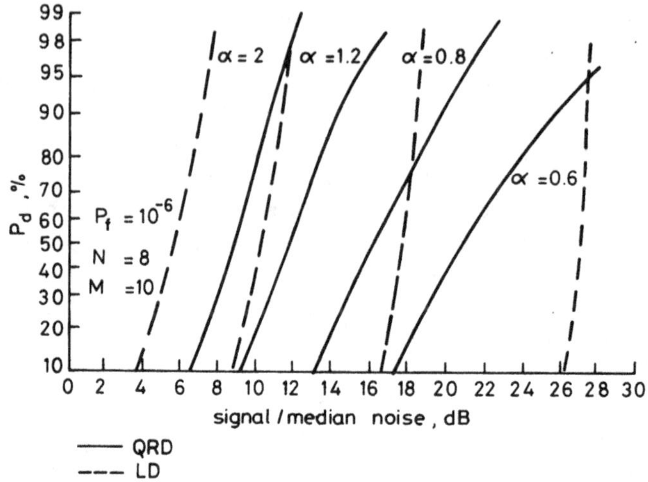

Fig. 3.34 *Detection probability versus signal-to-median clutter ratio*

for $\alpha = 1{\cdot}2$ and $0{\cdot}45N$ to N for $\alpha = 0{\cdot}8$. ORQT decreases as the number of pulses integrated M increases and approaches a steady value.

The detection probability P_d versus signal-to-median clutter ratio is shown in Fig. 3.34 for nonfluctuating signals at $P_{fa} = 10^{-6}$, $M = 10$, $n = 8$ and $\alpha = 2$, 1·2, 0·8 and 0·6. In computation the rank-quantisation threshold is taken to be asymptotic ORQT in Gaussian noise according to eqn. 3.140.

It can be seen from this Figure that the loss of QRD relative to the linear detector decreases with α, and can even become a gain. If a proper ORQT is taken for various α, M and N, the performance of the QRD may be better than the above.

Fang [39] studied the detection performance of a rank-sum (GS) nonparametric detector in Weibull clutter. Monte Carlo simulation combined with importance-sampling technique were used to simulate the detection performance of a rank-sum detector in Weibull clutter. The shape parameter of Weibull distribution was chosen as $\alpha = 1{\cdot}2$ and 0·6. The detection performance of a rank-sum (GS) nonparametric detector for a steady-state target signal is shown in Fig. 3.35*a* for $M = 10$, and Fig. 3.35*b* for $M = 30$, respectively. The detection performance for a log-normal fluctuating signal with mean-to-median ratio $\rho = 1{\cdot}2$ is shown in Fig. 3.36 for $M = 10$.

It can be seen from these Figures that the detection loss of rank-sum (GS) non-parametric detector in Weibull clutter depends on the shape parameter and number of reference cells. It increases with the decrease of the shape parameter α and the decrease in the number of reference cells. The detection loss for a fluctuating target is considerable, especially for the log-normal fluctuating target.

It is well known that the cell-average CFAR detector can maintain the false

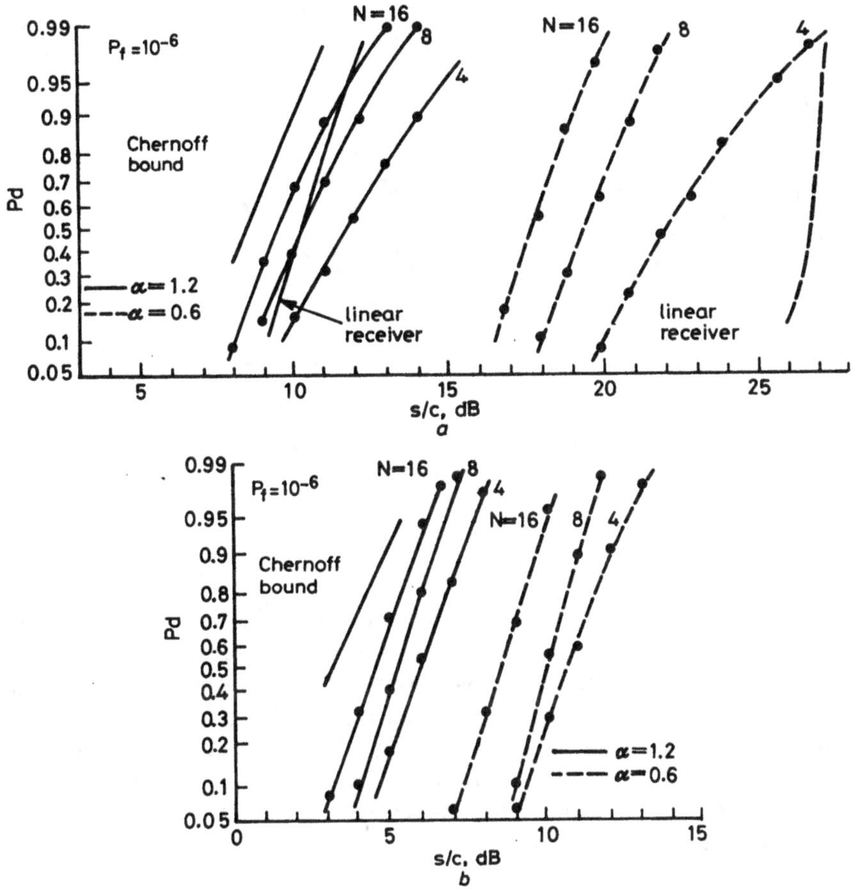

Fig. 3.35 *Detection performance of rank sum detector for steady-state target signal in Weibull clutter*
(*a*) $M = 10$
(*b*) $M = 30$

alarm constant in Rayleigh distribution. It is a parametric type of CFAR detector. When it is matched with the clutter distribution, i.e. if the clutter distribution can be fitted with Rayleigh PDF, it can be anticipated that the CFAR loss of the cell-average CFAR detector will be a minimum. Since most of the clutter can be fitted with Weibull distribution, it is of interest to compare the detection performance of a cell-average CFAR detector with that of a non-parametric CFAR detector in Weibull clutter.

Zhang *et al.* [40] compared the detection performance of a cell-average CFAR detector with that of a non-parametric CFAR detector, QRD and RSD, for steady-state and Swerling I and II targets in Weibull clutter. In the

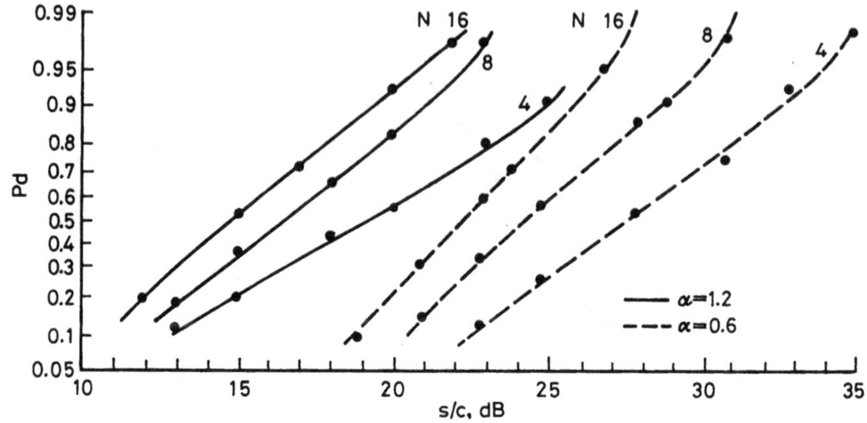

Fig. 3.36 *Detection performance for log-normal fluctuating target*
$P_f = 10^{-6}$
$M = 10$
$\rho = 1{\cdot}4$

simulation, three different shape parameters of Weibull-distribution clutter were used to represent the dynamic clutter environment.

The results for steady-state targets, Swerling I, and Swerling II, for both a cell-average CFAR detector and quantised-rank (QR) detector followed by a moving-window integrator are shown in Figs. 3.37*a*, *b* and *c*, respectively. The solid lines represent the detection performance of the QRD, and the dashed lines represent the detection performance of the cell-average detector. The false-alarm probability is fixed at 10^{-6}, and the number of pulse integrated is equal to 10.

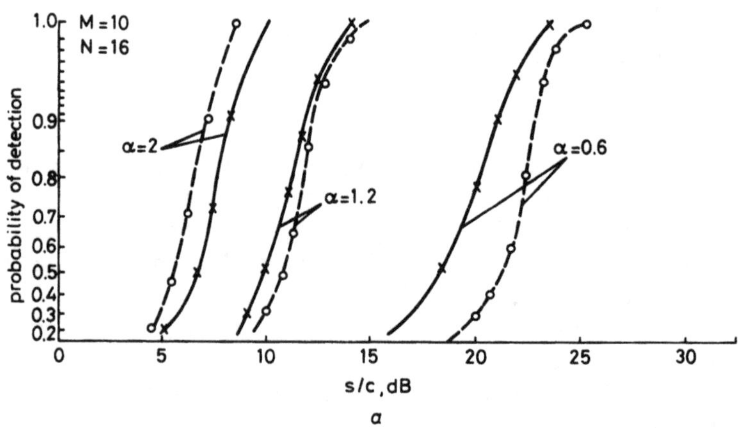

Fig. 3.37

(*Continued on opposite page*)

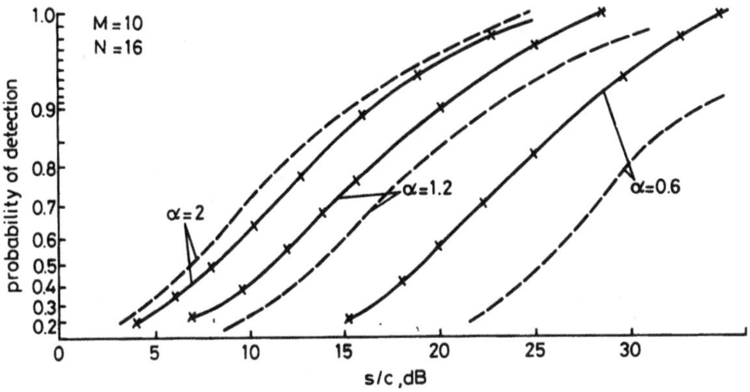

b

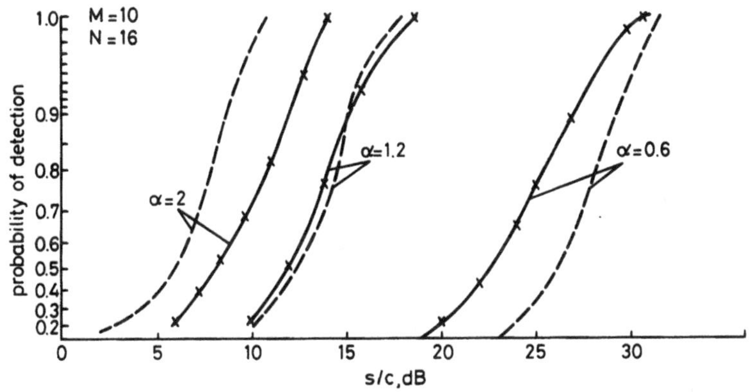

c

Fig. 3.37 *Comparison of detection performance of QRD and cell-average CFAR detector followed with a 10-pulse moving-window detector in Weibull clutter with different shape parameters*
(*a*) **Steady state target**
(*b*) **Swerling I target**
(*c*) **Swerling II target**

It can be seen from these Figures that the detection performance of a cell-average CFAR detector in Rayleigh clutter ($\alpha = 2$) is better than that of the quantised-rank detector. However, as the shape parameter α decreases to 1·2, its performance is nearly equal to that of the latter; and as the shape parameter α decreases to 0·6, the detection performance of the QRD is better than that of the former. The fluctuation loss for a Swerling I target is larger than that of a Swerling II target for both detectors.

To compare the detection performance of the quantised-rank (QR) detector and rank-sum (RS) detector with that of the cell-average (CA) detector, the

Table 3.17 *The nonparametric loss L_{np} in Weibull clutter for $M = 10$*

	n	4			8			16		
	α	2·0	1·2	0·6	2·0	1·2	0·6	2·0	1·2	0·6
RQ	Marcum	2·4	−2·5	−6·0	1·6	−1·9	−6·8	1·4	0·0	−1·6
	Swerlg. I	3·1	−2·9	−6·8	1·6	−3·0	−7·4	1·1	−2·8	−6·0
	Swerlg. II	4·8	1·5	−3·8	4·1	1·5	−4·4	3·4	0·1	−2·4
RS	Marcum	3·2	−1·7	−6·9	2·0	−2·4	−7·5	1·6	−0·8	−2·8
	Swerlg. I	4·2	−3·1	−7·0	2·5	−3·3	−7·8	1·8	−3·1	−6·4
	Swerlg. II	8·9	4·8	−4·2	5·2	1·7	−5·4	5·2	2·1	−5·4

Table 3.18 *The nonparametric loss L_{np} in Weibull clutter for $M = 30$*

	n	4			8			16		
	α	2·0	1·2	0·6	2·0	1·2	0·6	2·0	1·2	0·6
RQ	Marcum	0·4	−0·1	−1·1	0·1	−1·9	−2·3	0·6	−2·3	−3·4
	Swerlg. I	1·0	−0·8	−4·1	0·9	−1·1	−4·2	0·5	−2·5	−4·1
	Swerlg. II	1·7	0·2	−2·0	1·5	0·0	−0·4	0·8	−0·3	−3·0
RS	Marcum	0·5	−0·3	−1·3	0·3	−2·3	−3·0	0·3	−2·5	−3·0
	Swerlg. I	1·0	0·0	−4·0	0·6	−2·0	−3·0	0·5	−2·2	−3·5
	Swerlg. II	2·2	0·2	−2·0	1·3	−2·0	−2·6	0·8	−2·4	−3·1

non-parameter loss L_{np} is adopted. It is defined as the difference of required signal-to-clutter ratio (S/C in dB) between the non-parametric (NP) detector and the CA detector under the same condition of $P_{fa} = 10^{-6}$ and $P_d = 0{\cdot}9$. This non-parametric loss can be calculated using the following procedure.

Step 1: Using the importance-sampling theorem, all the combination values of the first and second threshold for each kind of CFAR detectors are calculated for a given false-alarm probability of 10^{-6}.
Step 2: The optimum first and second threshold are chosen to minimise the required S/C for different values of N and M, Weibull parameters, signal-fluctuation cases and signal powers under the condition of $P_d > 0{\cdot}5$.
Step 3: Subtract the required minimal S/C of the CA detector from the one of the NP detectors under the same condition of $P_{fa} = 10^{-6}$ and $P_d = 0{\cdot}9$; and L_{np} can be determined.

The results of the calculation are shown in Table 3.17 for $M = 10$ and Table 3.18 for $M = 30$.

Comparing the L_{np} of the QR detector with that of the RS detector, we find that the performance of the RS detector is worse than that of QR detector in Rayleigh clutter, but a little bit better than that of the QR detector in most Weibull-clutter situations especially of small value. When the number of integrated pulses is increased to 30, both detectors have similar performance.

It should be noted that all the results in this Section are based on the assumption of Independent and Identical Distribution (IID) of all the reference samples. This is true for most cases in the range direction but is not valid in the azimuth direction. The reason is that the clutter samples in the same range bin are correlated with each other within a radar resolution cell, which, in the azimuth direction, is nearly equal to the beamwidth, as mentioned before. When the pulses within the beamwidth are integrated, not only is the target signal enhanced but also the clutter signal. Therefore, the detection performance for all these different detectors is worse than that shown in the Figures. However, it is valid for the frequency-agility radar, which can decorrelate the clutter effectively. Apart from this, the relative performance comparisons are valid for most cases.

3.6 References

1 TRUNK, G. V., and GEORGE, S. F.: 'Detection of targets in non-Gaussian sea clutter, *IEEE Trans.*, 1970, **AES-6**, pp. 620–628
2 TRUNK, G. V.: 'Further results on the detection of targets in non-Gaussian sea clutter', *IEEE Trans.*, 1971, **AES-7**, pp. 553–556
3 TRUNK, G. V.: 'Detection of targets in non-Rayleigh sea clutter', IEEE EASCON Record 1971, pp 239–245

4 GOLDSTEIN, G. B.: 'False alarm regulation in log-normal and Weibull clutter', *IEEE Trans.*, 1973, **AES–9**, pp. 84–92

5 EKSTROM, J. L.: 'The detection of steady targets in Weibull clutter' *in* 'Radar present and future', IEE Conf. Publ. 105, London, Oct. 1973, pp. 221–226

6 SCHLEHER, D. C.: 'Radar detection in Weibull clutter', *IEEE Trans.*, 1976, **AES–12**, pp. 736–743

7 FARINA, A., RUSSO, A., and SCANNAPIECO, F.: 'Radar detection in coherent Weibull clutter', *IEEE Trans.*, 1987, **ASSP–35**, pp. 893–895

8 HANSEN, V. G.: 'Constant false alarm rate processing in search radars' *in* 'Radar present and future', IEE, Conf. Publ. 105, Oct. 1973, pp. 325–332

9 HANSEN, V. G.: 'Generalised constant false alarm rate processing and an application to the Weibull distribution'. 1972 International Symposium on Information Theory, Asilomar, Cal., USA

10 WEIBULL, W.: 'Moment estimators for Weibull parameters and their asymptotic efficiency', Lausanne, April 1969, AD 690162

11 MENON, M. V.: 'Estimation of the shape and scale parameters of the Weibull distribution', *Technometrics*, 1963, **5**, pp. 175–182

12 CRAMER, H.: 'Mathematical methods of statistics' (Princeton University Press, 1963)

13 SZAJNOWSKI, W. J.: 'Discrimination between log-normal and Weibull clutter', *IEEE Trans.*, 1977, **AES–13**, pp. 480–485

14 DUMONCEAUX, R., and ANTLE, C. E.: 'Discrimination between the log-normal and the Weibull distributions', *Technometrics*, 1973, **15**, pp. 923–926

15 TUGNAIT, J. K., and PRASAD, S.: 'Adaptive radar detection with asymptotically regulated false alarm rate', *IEEE Trans.*, 1977, **AES–13**, pp. 390–394

16 MARCUM, J. I.: 'A statistical theory of target detection by pulse radar-mathematical index', *IRE Trans.*, 1960, **IT–6**, pp. 59–267

17 VAN TREES, H. L.: 'Detection, estimation and modulation theory; Pt. 1 (Wiley, NY, 1968) Sec. 2.7

18 COLE, L. G., and CHEN, P. W.: 'Constant false alarm rate detector for a pulse radar in a maritime environment'. IEEE NAECON 78 Rec., 1978, pp. 1110–1113

19 BUCCIARELLI, T.: 'CFAR problems in Weibull clutter', *Electron. Lett.*, 1985, **21**, pp. 318–319

20 THOMAS, J. B.: 'Nonparametric detection', *Proc. IEEE*, 1970, **58**, pp. 623–631

21 DILLARD, G. M., and ANTONIAK, C.E.: 'A practical distribution-free detection procedure for multiple-range-bin radars', *IEEE Trans.*, 1970, **AES–6**, pp. 629–635

22 HANSEN, V. G., and OLSEN, B. A.: 'Non-parametric radar extraction using a generalized sign test', *IEEE Trans.*, 1971, **AES–7**, pp. 942–950

23 TRUNK, G. V., CANTRELL, B. H., and QUEEN, F. D.: 'Modified generalized sign test processor for 2-D radar', *IEEE Trans.*, 1974, **AES-10**, pp. 574–582

24 ZHU, Z. D.: 'The asymptotic performance of quantized rank non-parametric detector', *Acta Electronica Sinica*, 1980, (3), pp. 89–97

25 MAO, Y. H., ZHOU, Z. C., MENG, X. Y., XONG, F. Q., and ZHANG, S. Y.: 'A non-parametric CFAR detector implemented with CCD tapped delay line'. IEEE 1985 International Radar Conference, pp. 430–434

26 AKIMOV, P. S.: 'Non-parametric observed signal', *Radiotekhnika*, 1977, **32**, pp. 7–30 (in Russian)

27 SCHLEHER, D. C.: 'Radar detection in log-normal clutter'. IEEE Int. Radar Conf., April 1975, pp. 262–267

28 SCHWARTZ, M.: 'A coincidence procedure for signal detection', *IEEE Trans.*, 1956, **IT–2**

29 CONTE, E., IZZO, L., LONGO, M., and PAURA, L., 'Asymptotically optimum radar detectors in non-Rayleigh clutter', IEE Proc. Vol. 134, Pt. F, No. 7, Dec. 1987, pp. 667–672.

30 ZHU, Z. D.: 'Detection in Weibull and log-normal noise', *J. Electron.*, 1985, **2**, pp. 103–118

31 MODESTINO, J. W., and NINGO, A. Y.: 'Detection of weak signals in narrowband non-Gaussian noise', *IEEE Trans.*, 1979, **IT–25**, pp. 592–600

32 FEDELE, G., IZZO, L., and PAURA, L.: 'Optimum and suboptimum space-diversity detection in non-Gaussian noise', *IEEE Trans.*, 1984, **COM–32**, pp. 990–997

33 KUTOYANTS, Y. A.: 'On the asymptotic theory of signal detection in non-Gaussian noise', *Radio Eng. & Electronic Phys.*, 1976, **21**, pp. 74–81

34 IZZO, L., and PAURA, L.: 'Asymptotically optimum space diversity detection in non-Gaussian noise', *IEEE Trans.*, 1986, **COM–34**, pp. 97–103

35 CAPON, J.: 'Optimum coincidence procedures for detecting weak signals in noise'. IRE 1960 Int. Conv. Rec., Pt. 4, pp. 154–166

36 MCMILLAN, S., and STEWART, C.: 'A signal processor for a scanning marine radar', IEEE 1986 National Radar Conference, 1986, pp. 77–82

37 ZHU, Z. D., QIU, Z. M., and ZHANG, X. B.: 'Study on performance of nonparametric quantized rank detector'. CIE 1986 Int. Conf. on Radar, Nanjing, China, Nov. 1986, pp. 468–473

38 FANG, Z. G.: 'Detection performance of nonparametric radar rank detectors in Weibull clutter', *Acta Electronica Sinica*, 1982, (6), pp. 76–80

39 FANG, Z. G.: 'Simulation studies of radar detection performance in non-Rayleigh clutter', J. *Beijing Inst. Technol.*, 1984, (1), pp. 89–99

40 ZHANG, S. Y., MAO, Y. H., and FANG, Z. G.: 'The performance comparison between parametric and non-parametric CFAR detectors in Weibull clutter'. CIR 1986 Int. Conf. on Radar, Nanjing, China, 1986, pp. 456–461

Chapter 4

Suppression of Weibull clutter

4.1 Introduction

The suppression of radar clutter is a very important problem in modern radar technique, and is discussed in many papers and books. We are not going to discuss this problem in general, but will discuss some special problems concerned with Weibull clutter.

There are many ways to suppress radar clutter, and they can be summarised as follows:

(i) Preventing the clutter echoes from entering the radar antenna:
This method includes: surrounding the radar with a clutter shelter fence; installing the radar in a high mountain; tilting the radar antenna to a higher elevation angle. All these methods can be applied to existing radars.
(ii) Reducing the clutter energy by decreasing the size of resolution cell of the radar: This includes narrowing the beamwidth (it is often limited by the size of the antenna), narrowing the pulse width or employing pulse compression (this method is often used in marine radars to reject sea clutter).
(iii) Enhancing the signal-to-clutter ratio by shaping the beam pattern of the radar antenna: This includes reducing the elevation sidelobes on the lower side; and increasing the higher beam to enhance the signal strength of a close target.
(iv) Enhancing the signal-to-clutter ratio by employing the polarisation technique: Circular polarisation can reduce the equivalent RCS of a raindrop to 15—20 dB for most microwave radar, but at the same time reduce the RCS of an aircraft by only 5—7 dB. Therefore, more than 10 dB gain can be obtained. A similar technique for increasing the target-to-precipitation ratio is to use crossed linear polarisation (RCS is then σ_{HV} or σ_{VH}). While rain echoes are again reduced by 15—25 dB, the reduction in target RCS is generally greater than 7 dB. In general, the backscatter coefficient of sea clutter for horizontal polarisation is smaller than that for vertical polarisation for low-resolution radar.

(v) Preventing the receiver from saturation: STC is widely used for this purpose. Digital-controlled STC can be realised by adopting microwave PIN devices and programmable gain RF and IF amplifiers. Logarithmic amplifiers (IF or video) only can be used in non-coherent radars owing to the non-linearity.
(vi) Suppressing the clutter in the time domain: This includes employing a CFAR detector with adaptive threshold or clutter map. However, these methods can only produce super-clutter visibility. Frequency agility can decorrelate the clutter echoes effectively. This technique combined with video integration can be used to suppress clutter for non-coherent radars.
(vii) Suppressing the clutter in freqency domain: This method utilises the difference in Doppler frequencies between target and clutter, which is the most effective method of rejecting clutter. However, coherent radar or coherent-on-receive radar is needed for this method. With this method, sub-clutter visibility can be obtained.

It seems that all of these methods can be applied not only to Rayleigh-distributed clutter but also to Weibull-distributed clutter, since the only difference between Weibull distribution and Rayleigh distribution is the long tail. That is true in most cases, but a slight difference must be considered in realising of some of these techniques.

For example, when the resolution cell of a radar is reduced by reducing the pulse width, although the total energy reflected by the sea wave is reduced, the maximum amplitude of sea spikes remains unchanged. This means that the distribution of sea clutter changes from Rayleigh to Weibull. Fig. 4.1 shows the effect of reducing the pulse width of a vertical-polarisation X band radar from 400 ns to 40 ns [1].

In this case, if we retain the threshold constant, the false-alarm rate does not need to be decreased at the same time as expected, since the tail of the distribution is increased.

Another example is the polarisation properties of sea clutter. It is well known that the backscatter coefficients of horizontal polarisation are 10—20 dB smaller than those of the vertical polarisation at low frequencies, low

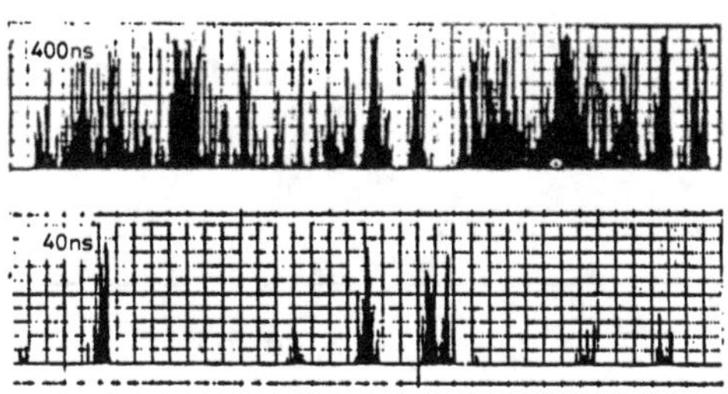

Fig. 4.1 *The effect on sea spikes of increasing range resolution (from Olin [1], © 1984 IEEE)*

grazing angles and low sea states. However, this difference tends to 0 dB at high sea states, high grazing angles and high frequencies. Furthermore, the amplitude distribution for horizontal polarisation is also different from vertical polarisation especially in high sea states and with high-resolution radar. Fig. 4.2 shows the amplitude distribution for both polarisation [1] (see also Fig. 2.17).

For comparison, a plot of Rayleigh-backscattered signal voltage (which is exponentially distributed in power or radar cross-section) is also shown. In the Figure the axes scales were chosen so that the Weibull distributions all plot as straight lines. It is clear that the distribution of the unmodified vertical and modified horizontal backscatter components fit the Weibull distribution. However, the unmodified horizontal deviates seriously from Weibull. It has a longer tail than the Weibull distribution: i.e., it is closer to log-normal distribution. A longer tail means that a higher threshold is needed to obtain the same false-alarm probability. This will result in a lower overall detection sensitivity to the desired target.

A long tail of the distribution function also means larger peak amplitude. This implies that a larger dynamic range is needed to avoid saturation of the radar receiver.

In this Chapter we will discuss some problems concerned with the suppression of Weibull clutter in time domain and frequency domain.

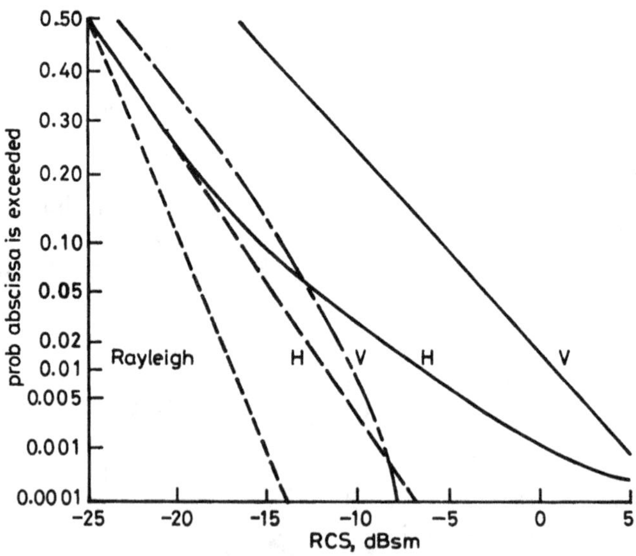

Fig. 4.2 *Horizontal- and vertical-polarisation amplitude distributions for unmodified (solid curves) and modified (broken curves) data of high-resolution (1°×40 ns) X-band radar at sea state 5 with grazing angle of 2·9° (from Olin [1], © 1984 IEEE)*

4.2 Suppression of Weibull clutter in time domain

Three types of time domain clutter suppression will be discussed:

(i) suppression of clutter within single sweep
(ii) suppression of clutter within single scan (multiple sweep)
(iii) suppression of clutter within multiple scan.

Although all these methods belong to the time domain, different features could be utilised, and we will discuss them in detail.

4.2.1 Suppression of clutter within a single sweep

This is the well known method of using a CFAR detector with adaptive threshold. Most of the adaptive thresholds are based on the estimation of the mean value of the clutter. It is evident that with this method, only super-clutter visibility can be obtained.

Brittain *et al.* [2] pointed out that, since the clutter distribution is closer to Weibull distribution than Rayleigh distribution, the detection threshold must be set appreciably higher than for a Rayleigh-distributed background for equivalent false-alarm rates. Therefore, additional loss will be introduced. This loss is called 'mean above clutter loss' (MACL) by Blythe and Treciokas [3]. In fact, it is caused by 'threshold above mean', since a margin should be added to the estimated mean value of the clutter to produce a threshold with a given false-alarm probability. Fig. 4.3 illustrates the mechanism of this loss.

It is of interest to study the super-clutter visibility that can be achieved with this method. Brittain *et al.* [2] calculated the detection probability versus target RCS under moderate ground clutter.

The radar parameters used to obtain the following results are 120 m range resolution. 2° beamwidth, 360° azimuth coverage, 10 s scan time and sensitivity of 110 dB signal-to-noise ratio at 1 km for 1 m^2 radar cross-section. The ground clutter is assumed to have log-normal distribution, and an average reflectivity of $-40\,dBm^2/m^2$ and an 84th percentile reflectivity of $-24\,dBm^2/m^2$.

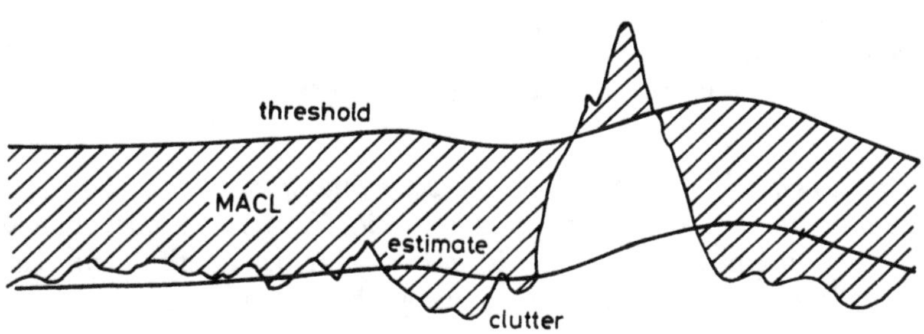

Fig. 4.3 *Threshold above mean loss (TAML)*

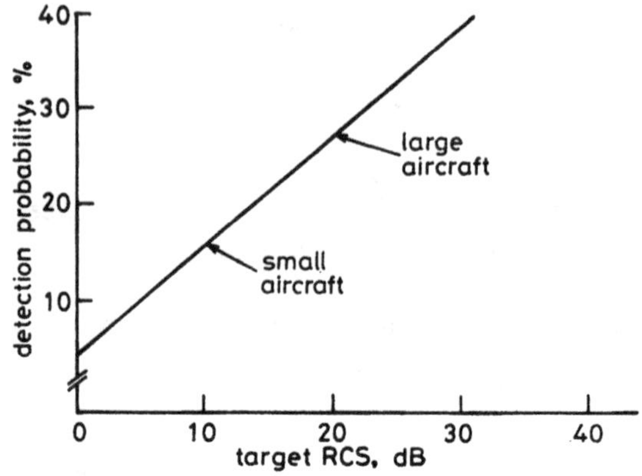

Fig. 4.4 *Detection probability versus target radar cross-section over moderate log-normal distributed ground clutter*
Note: Ground clutter log-normally distributed
$\sigma^0 = -40$ dBm²/m²
84th Percentile = −24 dBm²/m²
Detection statistics, target and clutter Rayleigh distributed

Simply by computing superclutter visibility from previously cited radar parameters, average clutter-reflectivity values with Rayleigh fluctuation, and a target range of 50 km, the average target detectability against its radar cross-section is shown in Fig. 4.4.

It is argued that radar cross-sections larger than average are involved since the zero-Doppler target is viewed at broadside.

Recognising that the Figure represents the best achievable detectability and is over moderate ground clutter, the performance is not impressive ($P_d = 34\%$ for a radar cross-section of 100 m²).

It is of interest to consider the superclutter visibility of this method in weather clutter. For the cited radar parameters at L-band, and a 2 mm/h rainfall rate, a superclutter of 15 dB is realised to 200 km for a 10 m² radar-cross-section (broadside) aircraft. This directly indicates that the superclutter visibility is high over an appreciable portion of the radar coverage in which light rainfall may occur.

For moderate rainfall rates (8 mm/h), 15 dB superclutter is achievable at 65 km. This is less important since the geographical extent is limited.

4.2.2 Suppression of clutter within single scan (multiple sweeps)

It is well known that the signal-to-noise ratio (SNR) can be enhanced by means of the integration of pulses within the radar beamwidth. The reason is that the target echoes in the same range bin of successive sweeps are correlated, but the samples of receiver noise of the same range bin but different sweeps are not correlated. Therefore, video integration gain against receiver noise can be obtained. It is approximately proportional to the square root of the number of pulses integrated.

However, this is not the situation for radar clutter. Since the clutter samples of the same range bin are correlated with each other, the effect of video integration for radar clutter is the same as the integration of target signals. In other words, the signal-to-clutter ratio (SCR) cannot be enhanced by video integration. If one would like to enhance the signal-to-clutter ratio by means of video integration, clutter-decorrelation measures should be employed.

It is well known that frequency agility is an effective method of decorrelating the clutter echoes. Since most of the clutter is the vector sum of the reflection echoes from a large number of small scatterers, the amplitude and phase of each component are related to the transmitting frequency of the radar. When the carrier frequency of the radar varies from pulse-to-pulse, the amplitude and phase of the resulting vector sum will also vary from pulse to pulse. This property is the frequency decorrelation of radar clutter.

Many authors have measured the frequency-correlation properties of different radar clutters.

Whitlock *et al.* [4] measured the frequency correlation of ground clutter with a C-band radar. The frequency range of the radar is between 5·35 and 5·85 GHz, and the pulse width is 2 μs. The result of measurement is shown in Fig. 4.5. The horizontal co-ordinate is the frequency difference of the adjacent pulse, and the vertical co-ordinate is the root-mean-square value of the

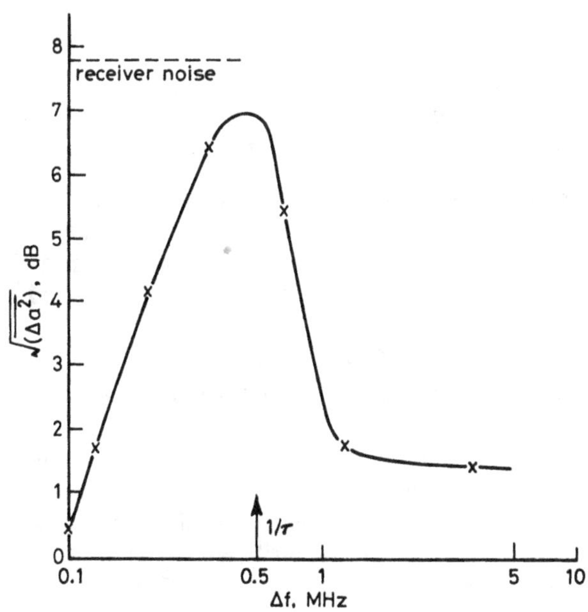

Fig. 4.5 *Measurement result of frequency correlation of C band ground clutter*

difference between two adjacent echo amplitudes with carrier-frequency difference Δf.

It can be seen from this Figure that, when the frequency difference between adjacent pulses is equal to the reciprocal of the pulse width, the RMS of the amplitude difference of adjacent pulses reaches maximum; i.e. the frequency-correlation coefficient is a minimum. When the frequency difference continuously increases, the RMS value of its echo-amplitude difference actually decreases; i.e. the frequency correlation coefficient becomes larger. This phenomenon is very difficult to explain; it may be caused by some practical problem in the experiment or result from the ground clutter itself.

Nathanson and Reilly [5] measured the frequency-correlation coefficient of rain clutter. It is defined as

$$\rho(\Delta f) = \frac{(\bar{I}_o)(\bar{I}) - (\bar{I})^2}{\bar{I^2} - (\bar{I})^2} \tag{4.1}$$

where I_o is the square of the signal amplitude at frequency f_o and I is the square of the amplitude at $f_o + \Delta f$. This function has been evaluated for the case in which the scattering volume contains many independent scatterers with more or less random positions. Under this condition the normalised frequency-correlation function of the echoes from multi-frequency rectangular pulses can be written as

$$\rho(\Delta f) = \left(\frac{\sin \pi\tau\ \Delta f}{\pi\tau\ \Delta f}\right) \tag{4.2}$$

where τ = pulse length
Δf = transmit-frequency change
$\rho(\Delta f)$ = normalised correlation coefficient (correlation of the echoes from two pulses transmitted with frequency difference Δf)

Fig. 4.6 shows Nathanson's measurement results, which coincide well with theory. The frequency shift Δf in all cases was 500 kHz, the pulses were approximately rectangular and of 0·4—3·2 μs duration, and the rainfall rate was high (20 mm/h). The extent of elevation of the illuminated area was about 400 m.

The experimental results conform closely to theory. This can be explained by the fact that the rainfall actually consists of a large number of small independent scatterer.

The situation for sea clutter is more complex. When the sea surface is illuminated with low-resolution radars, i.e. pulse length > 100 ns, the same results can be obtained. Fig. 4.7 shows the results measured by Pidgeon [6]. The experiments were performed primarily at C band (5·7 GHz) with both horizontal and vertical polarisation and with a 2.5° two-way beamwidth. The pulse width is varied from 0·1 to 10 μs.

Fig. 4.7 is a composite of the correlation coefficient versus $\tau\ \Delta f$ for the data.

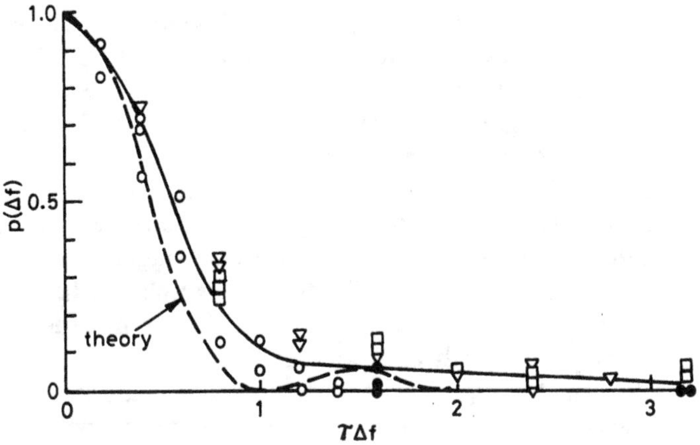

Fig. 4.6 *Correlation coefficient of rain echoes versus frequency shift-pulse length product (from Nathanson et al. [5], © 1968 IEEE)*
Beam diameter = 650 feet at gate
Each point = 1000 samples
● = 3·2 μs pulse (τ)
□ = 1·6 μs
▽ = 0·8 μs
○ = 0·4 μs
Rainfall rate = 20 mm/h
Frequency = 5·7 GHz

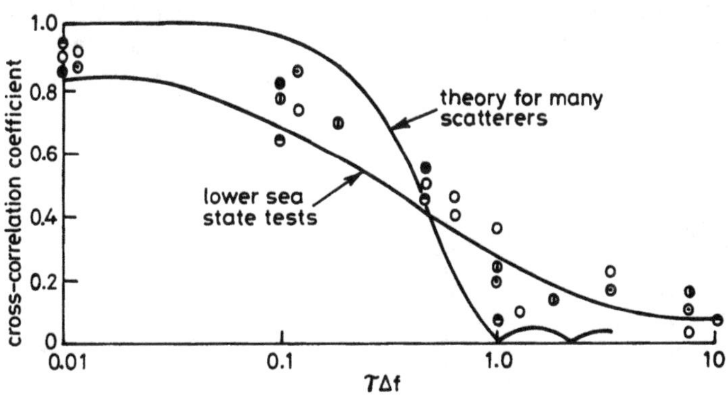

Fig. 4.7 *Frequency correlation of radar sea return at C band as a function of pulse length times frequency shift (vertical and horizontal polarisation) (from Nathanson [14], © 1969 McGraw Hill)*
Pulse width τ
○ 0·1 μs
⊙ 0·3 μs
● 1·0 μs
○ 3·0 μs
◓ 10 μs
0·2–4·0° depression angle
10–40-knot winds
1–8 ft waves

The upper curve is for an infinite collection of small scatterers. The experimental points for $\tau\ \Delta f \ll 1$ are less than unity away to the slight time decorrelation for signals with small frequency separations. The solid-line data represent points taken at about 10° grazing angle, wind speeds of 3—9 knots, and wave heights of 1/2 to 2·5 ft. The correlation coefficients at $\tau\ \Delta f > 1$ for pulse lengths of 0·1, 0·3 and 1·0 μs are all bellow 0·2, which indicates that the return is essentially decorrelated when considering the effects of receiver noise and finite sample lengths. The individual points are from higher sea-state tests.

Fig. 4.8 shows only the data points for the 0·1 μs pulse of horizontal- and vertical-polarisation transmissions at the lower sea state in order to observe the effect of shorter pulse length and polarisation.

It can be seen from this Figure that, although the spread in the computed correlation coefficient is somewhat greater for horizontal polarisation, the echoes seem decorrelated at $\tau\ \Delta f > 1$, However, it should be noted that, in this Figure, the mean backscatter power that is common to all frequencies for the 5 s computation period was subtracted before the correlation coefficient was computed. This means that an echo from an individual wave can be recognised by 100 ns pulse. This phenomenon becomes clearer for shorter pulse lengths. Ward et al. [7] showed two photographs obtained with a high-resolution radar (Figs. 4.9*a* and *b*).

These data were collected from a clifftop radar at I-band (8—10 GHz) incorporating frequency agility, with 1.2° beamwidth and 28 ns pulse length. Figs. 4.9*a* and *b* show range–time intensity plots of pulse-to-pulse clutter from a range window of 960 m at a range of 5 km and grazing angle of 1·5°. Fig. 4.9*a* is for fixed frequency and shows that, at any range, the return fluctuates

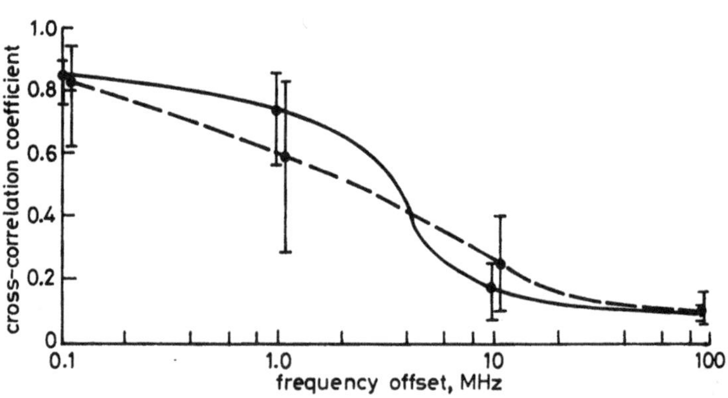

Fig. 4.8 *Frequency correlation of radar sea return for vertical and horizontal polarisation (pulse length* 100 *ns, C band) (from Nathanson [14], © 1969 McGraw Hill)*
Note: Bars I denote spread of data. Points through which curves are drawn are median values
● vertical polarisation
▲ horizontal polarisation

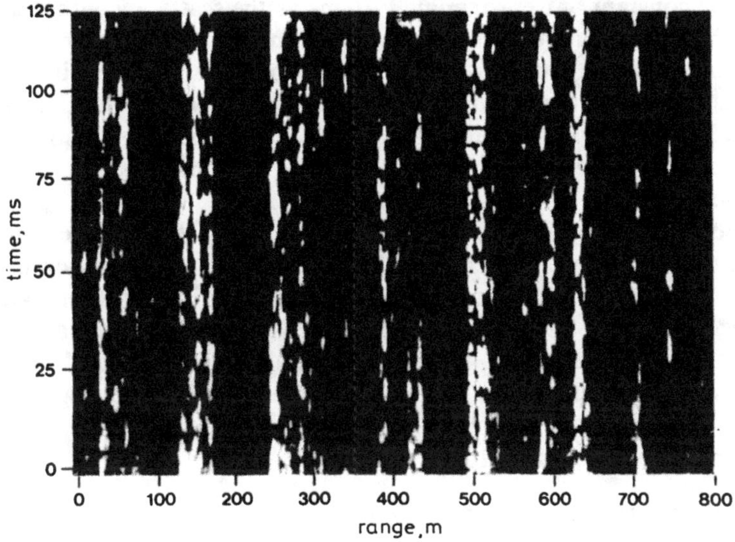

a

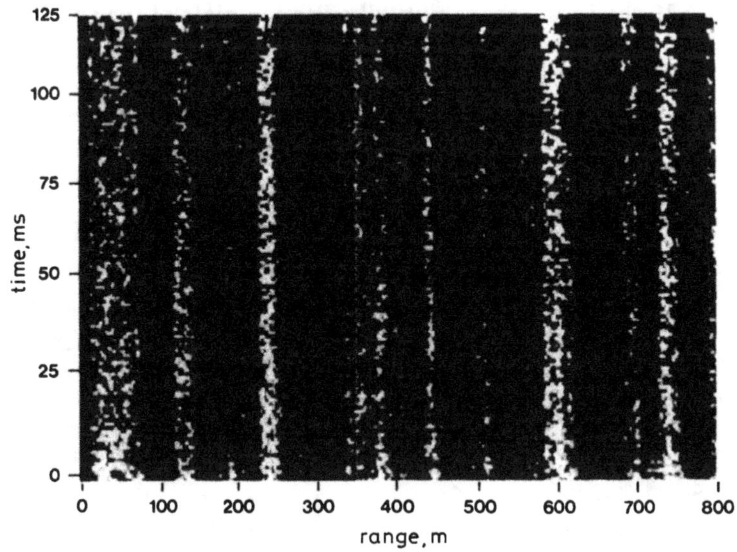

b

Fig. 4.9 *Sea-clutter record from fixed-frequency (a) and frequency-agility (b) high-resolution radar (Crown copyright/RSRE. Reproduced by permission of the Controller of HMSO.)*

with a time constant of approximately 10 ms as the scatterers within the patch move with the internal motion of the sea and change their phase relationships.

This very speckled pattern is decorrelated from pulse to pulse by frequency agility, as shown in Fig. 4.9*b*. Both figures show that the local mean level varies with range owing to bunching of the scatterers. This is unaffected by frequency agility. The total time (1/8 s) of Figs. 4.9*a* and *b* is not sufficient for the bunching change at any given range.

It has been found that the statistical distribution of the speckle component can be fitted with Rayleigh distribution (i.e. the central limit theorem applies within the cell) and the modulation fits the chi distribution (generalised for non-integer degrees of freedom). Therefore, Ward [8] suggested a compound form of the *K*-distribution. This model allows the temporal correlation properties of the clutter to be taken into account when assessing integrating detection schemes. Derivation of the amplitude distribution yields the *K*-distribution (see also eqn. 2.10)

$$p(x) = (2b/\Gamma(\nu))(bx/2)^{\nu} K_{\nu-1}(bx) \tag{4.3}$$

where x = amplitude,
$K_\nu(z)$ = modified Bessel function,
b = scale parameter
ν = shape parameter

Baker et al. [9] provided some experimental results obtained with the same radar. Fig. 4.10 shows the probability-distribution plots of the speckle component of sea clutter.

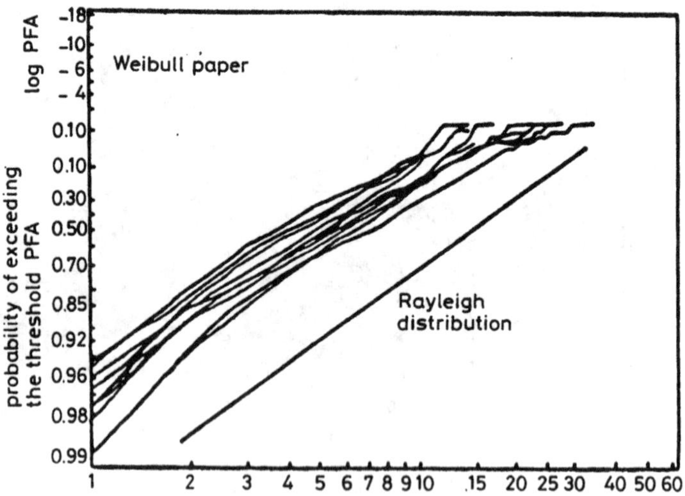

Fig. 4.10 *PDF of the speckle component of sea clutter*

As the sea clutter shows two dominant fluctuation components, a data set will generally contain many more independent values of the speckle than of the underlying modulation. The analysis has therefore been adapted to assume these two components, with a large number of independent speckle samples and a limited number of the modulation. Fig. 4.11 shows the bias and spread of third and fourth moment versus second moment which might be expected (± 3 standard deviations) for 100 independent values of the modulation [9].

The spread in expected moment values demonstrates the difficulty in matching models with limited experimental data sets.

The compound distribution of two components can be written as

$$p(a) = \int_{\nu x} p(x)p(a|x)\, dx \tag{4.4}$$

where $p(x)$ = PDF of the slow component (chi distribution)
$p(a|x)$ = PDF of the amplitude a, given a value of x (Rayleigh distribution of mean x)

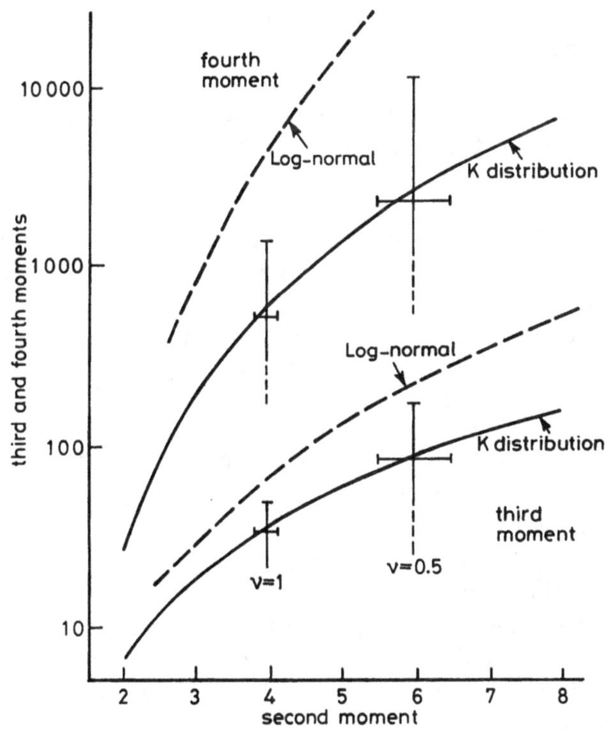

Fig. 4.11 *Expected spread and bias on normalised moments of the compound K-distribution, estimated from samples containing 100 independent values of the underlying modulation and an infinite number of values of speckle*

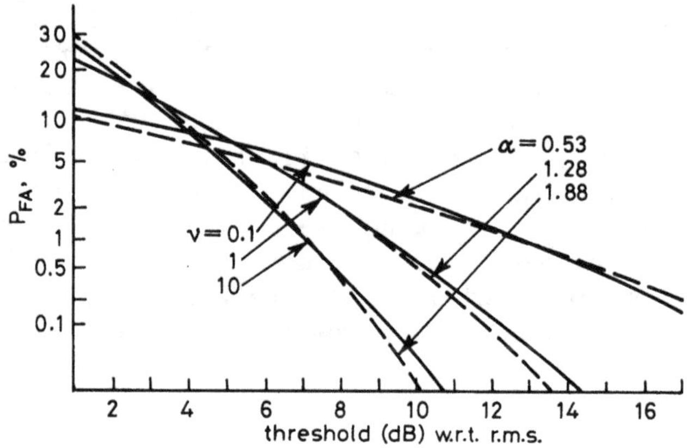

Fig. 4.12 *Cumulative Weibull and K-distributions*
α = Weibull parameter
ν = K parameter

Fig. 4.12 shows cumulative K and Weibull distributions plotted on log-normal paper. Each K-distribution ($\nu = 0{\cdot}1$, 1, 10) is matched to a Weibull using the first two moments. The horizontal axis is threshold with respect to RMS clutter.

It can be seen that Weibull and K-distributions are very similar and both have negative second differentials on this plot. This implies less of a 'tail' to the distributions than log normal (a straight line on this paper). Perhaps more important is the relationship of the curves within a family. At high thresholds the probability of false alarm increases with the higher moments (as ν and α decrease). This is expected from the concept of 'spikiness'. However, at lower thresholds the trend reverses (as it must, since the mean powers are the same). This has consequences for single-hit and binary-integration detection.

Fig. 4.13 shows the single-hit detection performance for a non-fluctuating target in K-distribution clutter. As expected, for low probability of false alarm, the signal-to-clutter ratio (SCR) increases as the value decreases and the clutter becomes more spiky. For high P_{fa} ($0{\cdot}1$), the SCR required decreases for 90% probability of detection P_d, and wavers for 50% P_d. It can be expected that the single-hit detection performance in Weibull clutter will be very similar to these plots.

For any fixed threshold system of processing where the integration is short compared to the correlation time of x, performance can be calculated assuming that x is constant and $y = a/x$ is independent from pulse to pulse.

This leads to the expression for probability of false alarm, P_{fa}, for a linear analogue integrator,

$$P_{fa} = \int_{\nu x} p(x) P\left(\sum_i y_i > t/x\right) dx \qquad (4.5)$$

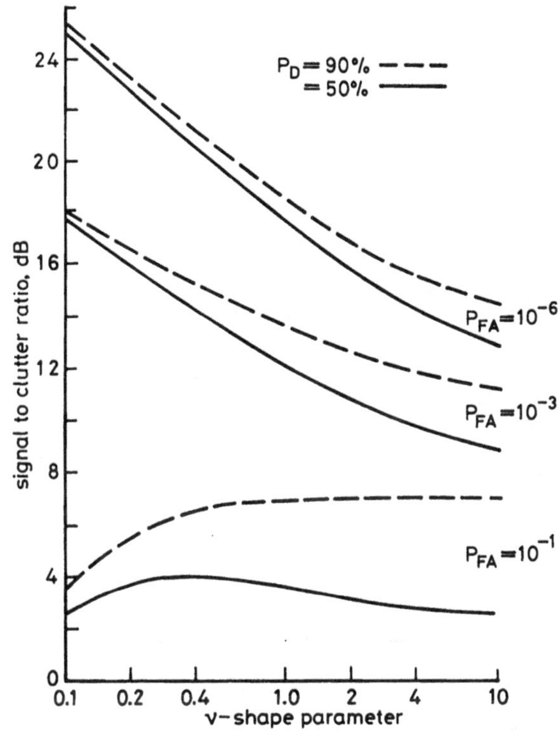

Fig. 4.13 *Single-hit detection performance for non-fluctuating target in K-distributed clutter*

where t = threshold

$P\left(\sum_i y_i > t/x\right)$ = probability of the test statics, $\sum_i y_i$, being greater than t/x, assuming independent y_i

A similar formula can be obtained for the probability of detection and also for other processing schemes. The resulting performance will include pulse-to-pulse correlation effects. Fig. 4.14 shows the results obtained by numerical computation for analogue and binary integration of 10 returns. The measure used for performance is the signal-to-RMS clutter ratio required for 50% probability of detection at a given false-alarm rate. This differs from the signal-to-median clutter ratio quoted elsewhere, and is chosen since RMS clutter can be directly related to σ_o, the usual clutter-level measure. The difference between median and RMS is not insignificant; for $\nu = 0{\cdot}1$, the RMS-to-median ratio is 25 dB.

It is of interest to compare the results of detection performance in *K*-distribution obtained by Ward [8], and in Weibull distribution by Schleher (Fig. 3.25*b*). The signal-to-median clutter ratio required for $P_d = 50\%$ is equal to

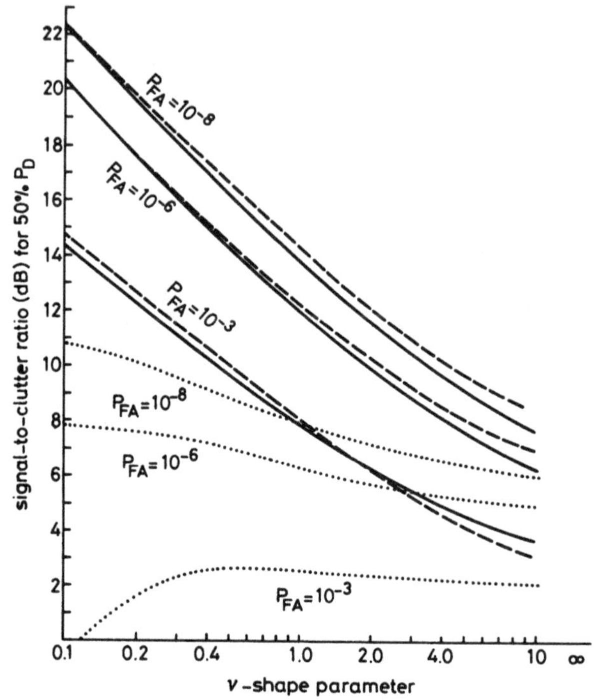

Fig. 4.14 *Performance for binary and analogue integration of 10 non-fluctuating returns in K-distributed clutter*
---- binary
—— analogue
······· binary (independent)

8·5 dB in Weibull clutter with $\alpha = 1{\cdot}2$, which is equivalent to $\nu = 1{\cdot}0$ from Fig. 4.11. The required signal-to-RMS clutter for the same condition in K-distributed clutter is about 11·5 dB. Unfortunately, the difference between signal-to-median ratio and signal-to-RMS clutter ratio at $\nu = 1{\cdot}0$ has not been given. One can believe that the results shown in Fig. 4.14 are more accurate, since the correlation between echoes has been considered. Because the K-distribution is very similar to Weibull distribution, the results shown in Fig. 4.14 can also be used for Weibull distribution.

4.2.3 Suppression of clutter within multiple scans

This technique, sometimes called clutter-map CFAR detection, for clutter suppression is somewhat similar to clutter suppression based on the adaptive threshold of CFAR detector. The adaptive threshold of CFAR detector is obtained by averaging the outputs of nearby reference cells, while in clutter-map CFAR detection the output of each resolution cell is averaged over several scans in order to obtain the background estimate. The former technique works well only if the background clutter is statistically homogeneous over range direction. The latter technique is preferred when the background is not homogeneous, which is often the case for practical environments in the real world.

The former technique is implemented using a moving-window integrator. Similarly, the clutter-map CFAR technique could use a moving-window estimator over several prior scans for each resolution cell.

As the average process takes a very long time (several scans), and the clutter background might be a time varying non-stationary statistics, the latter techniques often use a background estimate derived by exponential smoothing of the output of each resolution cell.

The CFAR loss-of-clutter map technique is analysed. Unfortunately, it is assumed that the predetector statistics of the complex envelope are complex Gaussian for the sum of thermal noise plus clutter plus target, for the purpose of simplifying the computation.

However, just as we mentioned before, this clutter-map adaptive thresholding only possesses super-clutter visibility. The purpose of employing clutter map is to maintain the false-alarm constant rather than to suppress the clutter itself.

4.3 Suppression of Weibull clutter in frequency domain

It is well known that clutter suppression in frequency domain, i.e. utilising Doppler frequency difference between clutter and target signal, is the most effective way to reject clutter. With this method, sub-clutter visibility can be obtained. This means that the target signal can be detected under very low SCR, e.g. -40 to -60 dB.

The most often used techniques are the wellknown MTI and MTD techniques. More than several hundreds papers have dealt with these topics. Unfortunately, most used the Gaussian clutter model not only for amplitude distribution but also for the shape of the clutter spectrum. The main reason is that the Gaussian model is easier to analyse and compute, but this model is not always true for real clutter. As mentioned above, in many cases the amplitude distribution of clutter can be fitted with Weibull distribution, and the spectrum of clutter can be expressed with AR spectrum.

Schleher [10] analysed the detection performance of non-recursive MTI filters in Rayleigh and log-normal clutter. He found that MTI performance in log-normal clutter is degraded from that available in Rayleigh clutter. For example, a correlated log-normal clutter with parameters $\sigma = 1$ and $\rho = 0{\cdot}95$, where σ and ρ are the standard deviation and correlation of the underlying two-dimensional Gaussian distribution that generates the correlated log-normal process after passage through an exponential nonlinearity, is compared against Rayleigh clutter. If the $P_{fa} = 10^{-6}$, the threshold for Rayleigh clutter can be determined to be 9·73 dB, while the threshold determined for log-normal clutter is 66 dB indicating severe performance degradation.

There are several reasons why large degradation can be expected. First, linear MTI is not optimum for processing log-normal clutter; nonlinear processors would provide better performance. Secondly, the comparison is unfair in the sense that the spectrum of the log-normal clutter process is spread with respect to the spectrum of Rayleigh clutter, thereby making the linear MTI filter less effective in reducing clutter.

The detection performance of the MTI filter in Rayleigh clutter can be conveniently determined using the concept of detection loss, which is given by [11]

$$L_R = (P_c/I\sigma_n^2 + 1) \tag{4.6}$$

where P_c is the clutter power, σ_n^2 is the noise power, and I is the MTI improvement factor. The detection loss is defined as the additional signal-to-noise ratio required in Rayleigh clutter to achieve a specified performance level (P_d, P_{fa}) relative to that provided by a standard square-law detector in receiver noise. When the MTI is operated in log-normal clutter, an additional detection loss is incurred. In Ref 11, the MTI detection loss in log-normal clutter, relative to that in Rayleigh clutter, was evaluated. The results for several typical situations are depicted in Figs. 4.15 and 4.16.

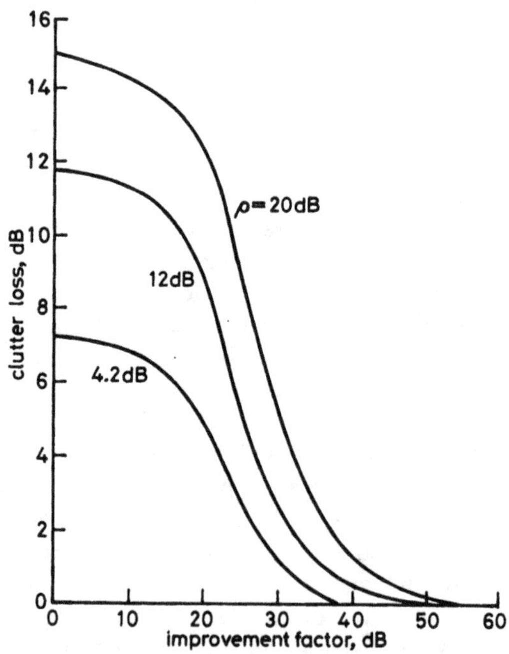

Fig. 4.15 *Detection loss as function of MTI improvement factor*
$P_d = 0{\cdot}9$; $P_{fa} = 10^{-6}$; $C/N = 20$ dB; $\rho = \exp \sigma_v^2/2$

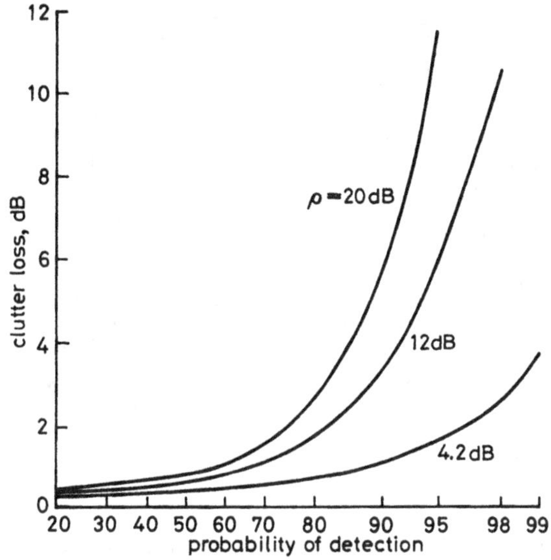

Fig. 4.16 *Detection loss as function of probability of detection*
$P_{fa} = 10^{-6}$; $C/N = 20$ dB; $I = 30$ dB, $\rho = \exp \sigma_v^2/2$

In Fig. 4.15, the detection loss is given as a function of MTI improvement factor for various values of clutter mean–median ratios ρ.

The detection loss is most severe for low to moderate values of the MTI improvement factor and for the more highly skewed clutter distribution.

Fig. 4.16 gives the detection loss as a function of probability of detection for a constant value of improvement factor (30 dB). Here the detection loss increases more rapidly for values of the detection probability above 0·8.

In general, the detection-loss curves indicate the value of designing MTI radars with large improvement factors when confronted with log-normal clutter. If large improvement factors are not achieved, the losses become excessive and only modest detection probabilities can be achieved.

Although these results were obtained for the log-normal distributed clutter, the conclusion is also suitable for Weibull distributed clutter, since the common feature of these two distributions is the long tail relative to Rayleigh distribution. It can be expected that the degradation of detection performance of the non-recursive MTI filter in Weibull clutter will be less than that of MTI filter in log-normal distributed clutter.

Farina *et al.* [12, 13] studied the problem of coherent detection in Weibull clutter. For the purpose of computer simulation, a coherent Weibull random sequence was generated. This sequence has a Weibull PDF for the amplitude, a unform PDF for the phase and an ACF (autocorrelation function), between the successive samples, selected at will. The generation of the Weibull-distributed sequence is shown schematically in Fig. 4.17.

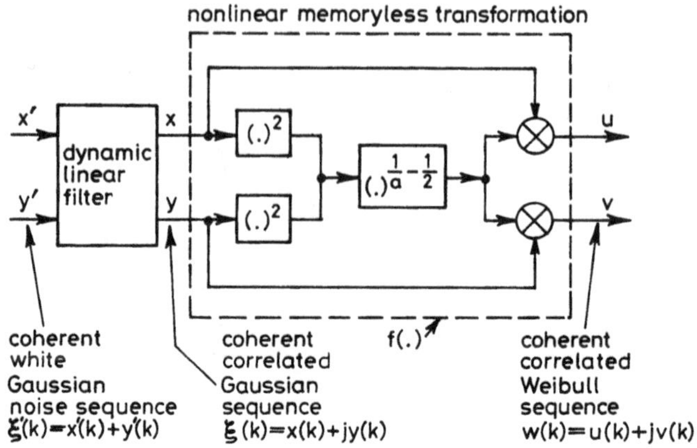

Fig. 4.17 *Generation of coherent Weibull distributed clutter*

A coherent zero-mean WGN (white Gaussian noise) sequence of N samples in time, $\{\xi'(k) = X'(k) + jY'(k), k = 1, 2, \ldots, N\}$, feeds the cascade of a linear dynamic filter and a nonlinear memoryless transformation $f(\cdot)$. The linear filter introduces a proper correlation between the samples of the sequence, i.e. the spectral width of the clutter. The bandlimited coherent Gaussian variable $\xi(k) = X(k) + jY(k)$ obtained at the output is transformed into the coherent Weibull variable $W(k) = U(k) + jV(k)$ by the nonlinearity $f(\cdot)$.

The linear dynamic filter is an FIR filter with $(N-1)$ taps, N being the length of the sequence to be generated. The weights of the FIR filter are chosen to control the covariance matrix (i.e. the process frequency spectrum) of the sequence. A mathematical relationship has been found between the correlation coefficient p of any two samples of Z and the correlation coefficient q between the corresponding samples of W:

$$q = K_a p + (1 - K_a) \tag{4.7}$$

where K_a depends on the skewness parameter a. A set of K values is

$$K_{0\cdot6} = 1{\cdot}758, \quad K_{0\cdot8} = 1{\cdot}406, \quad K_{1\cdot2} = 1{\cdot}112$$

Eqn. 4.7 has been checked with successes by resorting to a Monte Carlo simulation on a digital computer.

The nonlinear transform is equivalent to multiplying the real-valued Weibull variate $((X + Y)^{**}(1/a - 1/2))$ by the function $\exp(j\varphi)$, where $\varphi = \text{arctg}(Y/X)$ is evenly distributed in $[0, 2\pi]$.

The Weibull PDF depends on two parameters: the skewness parameter a and the scale factor b. When $a = 2$, the PDFs of U and V are Gaussian and the corresponding PDF of the amplitude is Rayleigh. When $a = 1$, an exponential PDF of the amplitude is obtained. As a tends to zero, the tails of the PDFs of U, V and $(U^2 + V^2)^{1/2}$ grow up. The PDF of arctg (U/V) is uniform

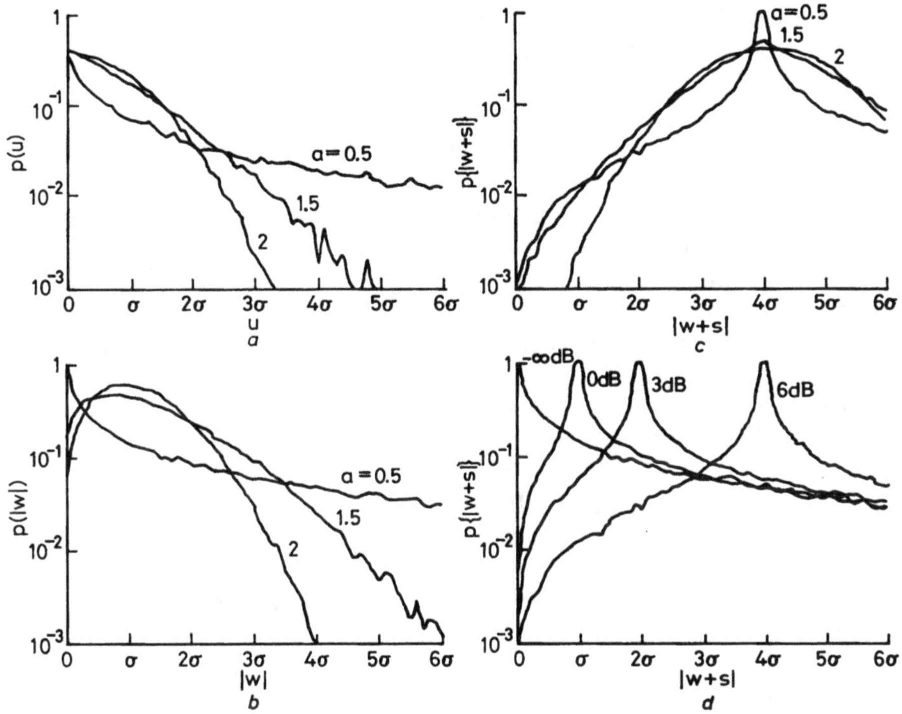

Fig. 4.18 *Histogram of computer-generated coherent Weibull clutter*
(*a*) In-phase (or quadrature) component
(*b*) Amplitude distribution
(*c*) Signal + Weibull clutter (SCR = 6 dB) with skewness *a* as parameter
(*d*) Signal + Weibull clutter (skewness *a* = 0·5, SCR as parameter)

in $[0, 2\pi]$. The scale factor b of the Weibull distribution is related to the power of the Gaussian sequence Z by the relationship $E(Z^2) = b^a$.

Some results of computer-generated histograms of Weibull-distributed clutter and clutter plus signals with this method are shown in Figs. 4.18*a*—*d*.

With this nonlinear memoryless transformation, the inverse transform can be realised; i.e. the input Weibull clutter can be transformed to Gaussian distribution. The only difference is that the real-valued Gaussian variate is obtained by $[(U^2 + V^2)^{**}(a/4 - 1/2)]$, where U and W are the inphase and quadrature components of the input Weibull variate.

Based on this principle, a nonlinear prediction filter was proposed by Farina et al. [12], the process being 'Gaussian-whitening-Weibulling'. This means that the input Weibull clutter is first transformed to Gaussian variate, then is whitened with an ordinary linear prediction filter, and finally is transformed back to Weibull distribution. However, the skewness of the input Weibull sequence must be known *a priori*, otherwise the transformation cannot be realised.

With this nonlinear prediction filter, a new family of detection processors for detection of target signals embedded in Weibull clutter was proposed by Farina et al. [12].

4.3.1 Detector for target signal known a priori embedded in Weibull clutter

In this case a target signal is known *a priori* embedded in coherent Weibull clutter with known skewness parameter and white Gaussian noise. The detector for this case is shown in Fig. 4.19.

The received echo $z(k)$ is processed through two nonlinear prediction filters. The upper filter is matched to the condition that the signal to be detected is the sum of target plus clutter (channel H_1), while the lower filter is constructed on the condition that the signal to be detected is just the clutter source (channel H_0). Channel H_1 differs from channel H_0 for the presence of target signal $s(k)$ is known *a priori*. The signal $s(k)$ is first subtracted from the incoming radar echo $z(k)$, so that the estimation of the disturbance $d(k)$ can be performed as in the channel H_0, and hence summed again downstream from the nonlinear prediction filter. By making the difference of the two estimates $\hat{z}_i(k/k-1)$ and $\hat{z}_0(k/k-1)$, so far obtained with the incoming echo, $z(k)$, two residuals, v_j and v_o, of the estimates are obtained. To ascertain which of the two residuals is zero-mean WGN of the train. The result is compared with a suitable threshold to obtain the desired P_{fa} value and maximise P_d. The two parallel nonlinear estimators, matched to the two alternative hypotheses, can be regarded as a means of obtaining a zero-mean white Gaussian sequence along that channel corresponding to the hypothesis which holds at present.

Although analytical evaluations are possible for the case of skewness parameter $a = 2$ and for any number of pulses N, or for $N = 2$, and for a not equal to 2, it is difficult to evaluate the detection performance for other cases. To avoid the limitations of the analytical method, Monte Carlo simulation was performed with a digital computer. The false-alarm probability has been set to 10^{-4} to limit the number of independent trials to be made. In doing the simulations, the Weibull-clutter spectrum has been assumed as having a Gaussian shape with a zero-mean Doppler frequency, i.e. $F_c = 0$. The spectrum depends on two parameters, namely: (i) the clutter/noise power ratio (CNR) and (ii) the autocorrelation coefficient q between any two couples of contiguous clutter samples. Additionally, the target signal has been assumed to have a Doppler frequency F_s equal to 0·5PRF and an assigned signal/noise power ratio (SNR) value.

The curves of Figs. 4.20*a*, *b* and *c* are related to the case of $N = 2$ echoes in the train. In particular, Fig. 4.20*a* is concerned with the case of skewness parameter $a = 1{\cdot}2$ and different values of the one-step autocorrelation coefficient q of the clutter. It is seen that an increase in the parameter q produces an improvement in the detection performance. This can be explained through a better estimation of the clutter portion in the received echo, which results in better cancellation of the disturbance. Fig. 4.20*b* refers to a lower value of a

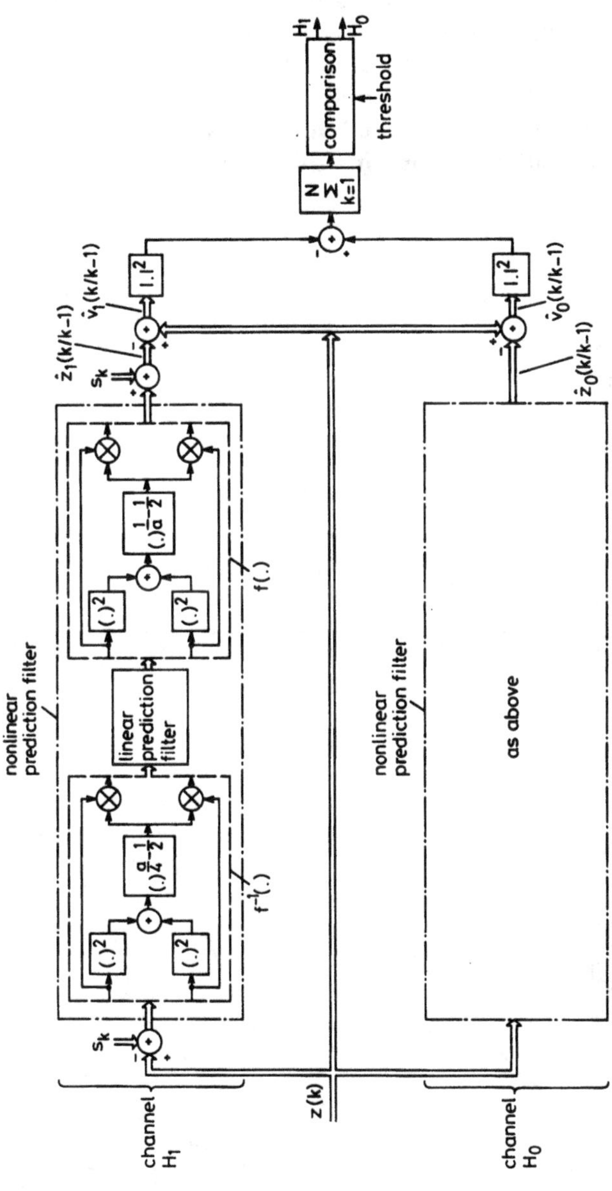

Fig. 4.19 *Detector for target signal known a priori embedded in Weibull clutter with known skewness and white Gaussian noise*

(i.e. $a = 0{\cdot}6$) which corresponds to a highly skewed clutter. Comparison with the previous set of curves shows a penalty in terms of detection performance owing to the longer tail of the clutter. This concept is better expressed by Fig. 4.20*c*, which refers to a specified value of the autocorrelation coefficient ($q = 0{\cdot}95$) and different values of skewness parameter *a*.

Figs. 4.21*a*—*d* show the detection performance for $N = 3$ under the same conditions. In particular, Fig. 4.21*a* illustrates the detection performance for the skewness parameter $a = 1{\cdot}2$ and having as parameter the correlation coefficient *q* of the clutter. Comparing this figure with Fig. 4.20*a*, it is noted

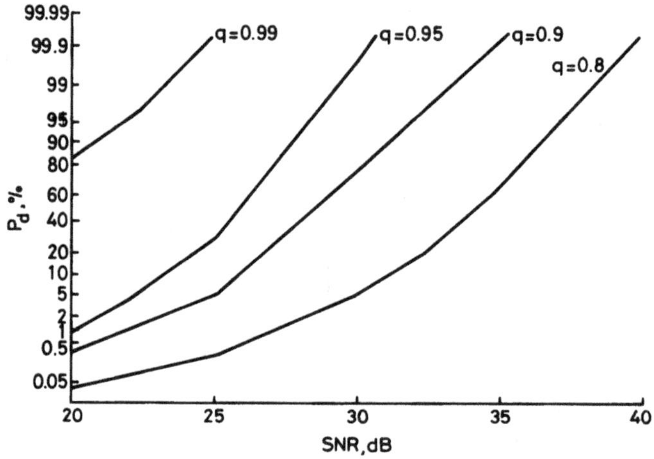

a

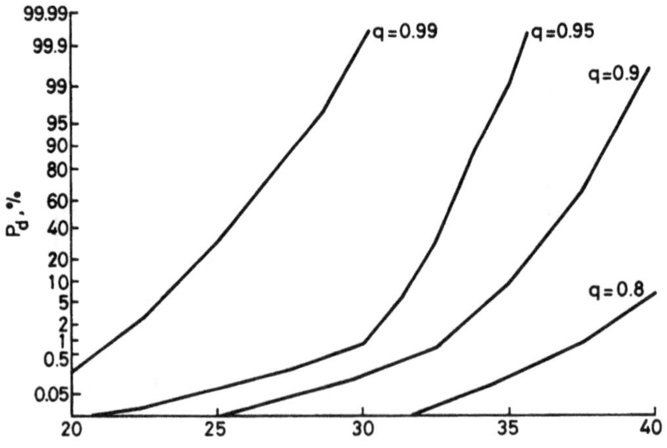

b

Fig. 4.20

(*Continued on next page*)

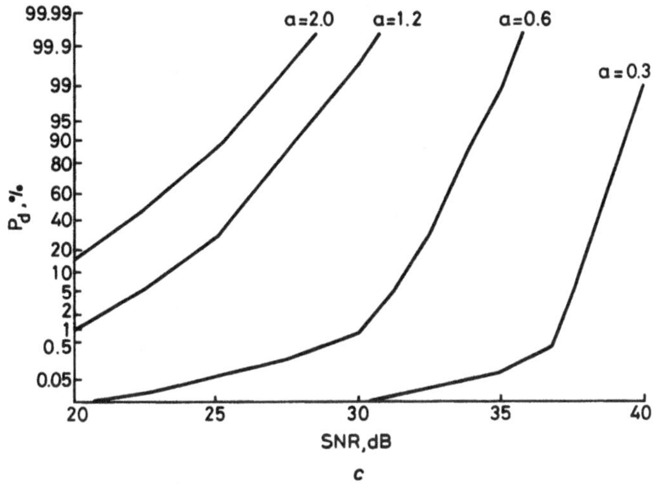

Fig. 4.20 *Detection performance of a target known a priori in coherent Weibull clutter*
$N=2$, $P_{fa}=10^{-4}$, CNR = 30 dB, $F_c=0$, $F_s=0{\cdot}5$ PRF
(*a*) $a=1{\cdot}2$, q as parameter
(*b*) $a=0{\cdot}6$, q as parameter
(*c*) $q=0{\cdot}95$, a as parameter

that a reduction in SNR of about 10 dB or more, on average, is obtained by increasing the number of pulses from two or three. Fig. 4.21*b* similarly corresponds to Fig. 4.20*b*, and Fig. 4.21*c* is similar to Fig. 4.20*c*. Again, a comparison of Fig. 4.21*c* with Fig. 4.20*c* shows the saving of SNR by processing three pulses in lieu of two. Fig. 4.21*d* shows the detection performance for $P_{fa} = 10^{-6}$ and $N = 3$. These curves should be compared with those of Fig. 4.21*a* to obtain a feeling for the SNR increases owing to the very low value of P_{fa}.

Since in most case the skewness of Weibull clutter is unknown, it is of interest to assess the robustness of the processor matched to the Gaussian case (i.e. $a = 2$) when fed with Weibull clutter. Fig. 4.22 shows the detection loss suffered by this processor matched to the Gaussian-clutter case when it is fed with Weibull clutter.

It is seen that a loss of 2 dB is suffered when $P_d = 0.9$ and the skewness parameter $a = 1{\cdot}2$. The loss rises to 4 dB when the parameter is equal to 0·6; and rises to 7 dB for a P_d of 0.5.

Figs. 4.23*a* and *b* show the detection performance and detection loss due to mismatching for $N = 4$.

Comparison of the curves in Fig. 4.23*a* with the companion curves of Figs 4.20*a* and 4.21*a* shows the SNR saving when the number of processed pulses increases. It can be seen from Fig. 4.23*b* that, for $P_d = 0{\cdot}9$, the loss is negligible when $a = 1{\cdot}2$ while it is of the order of 2 dB when $a = 0{\cdot}6$. The

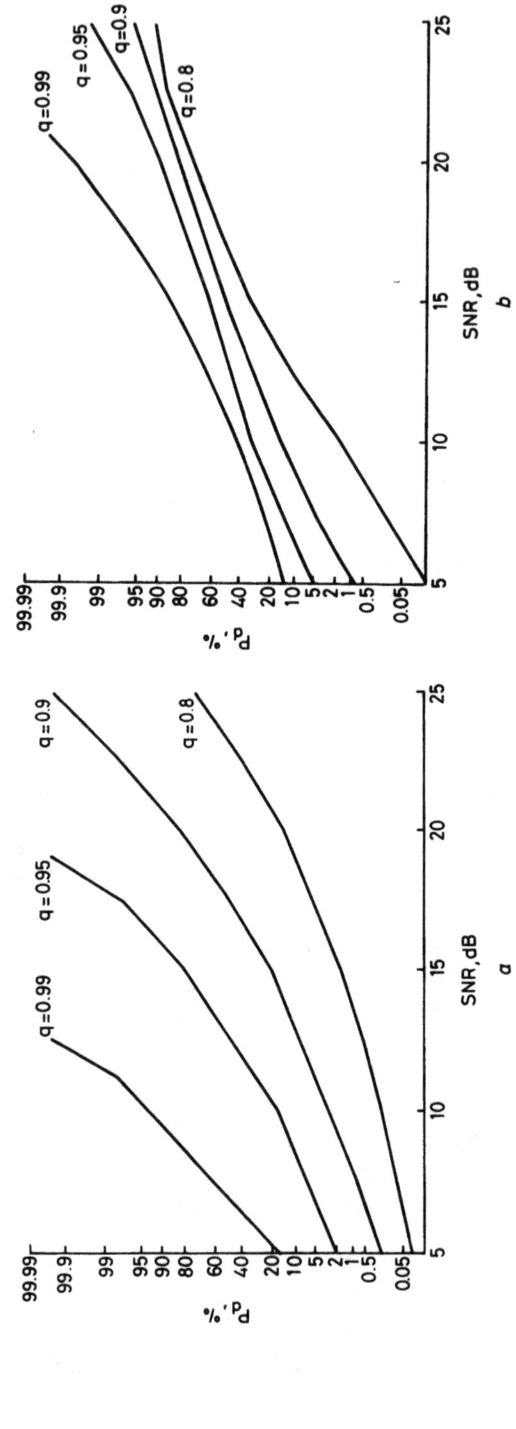

Pd, %
99.99
99.9
99
95
90
80
60
40
20
10
5
2
1
0.5
0.05
q=0.99
q=0.95
q=0.9
q=0.8
5
10
15
20
25
SNR, dB
a
b

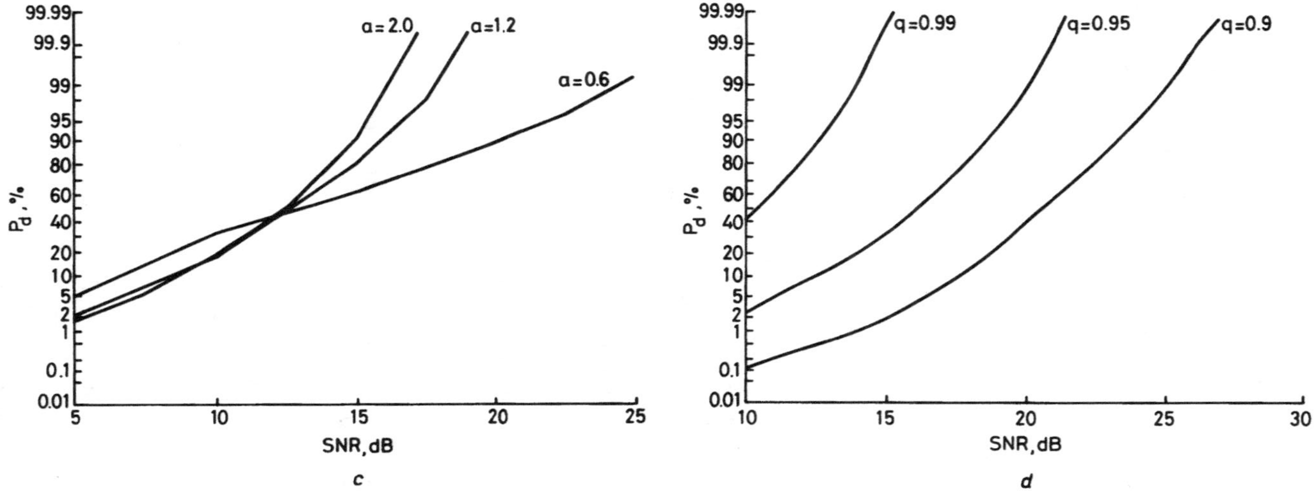

Fig. 4.21 *Detection performance of a target known a priori in coherent Weibull clutter* $N=3$, $P_{fa}=10^{-4}$, CNR = 30 dB, $F_c=0$, $F_s=0{\cdot}5$ PRF

(*a*) $a=1{\cdot}2$, q as parameter
(*b*) $a=0{\cdot}6$, q as parameter
(*c*) $q=0{\cdot}95$, a as parameter
(*d*) $a=1{\cdot}2$, q as parameter and $P_{fa}=10^{-6}$

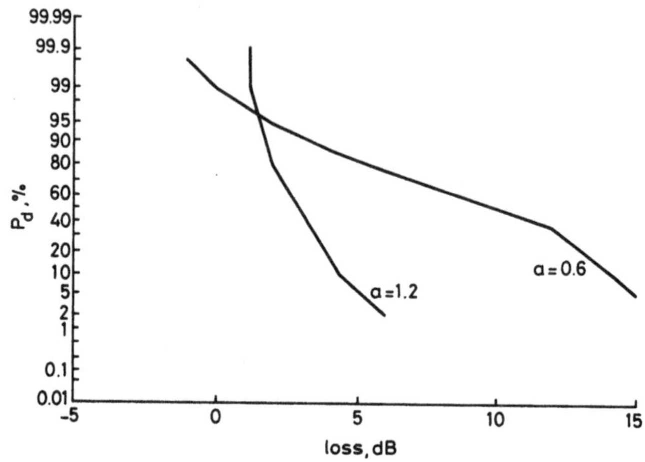

Fig. 4.22 *Detection loss due to mismatching*
$N=3$, $q=0{\cdot}95$, $P_{fa}=10^{-4}$, CNR = 30 dB, $F_c=0$, $F_s=0{\cdot}5$ PRF, with *a* as parameter

losses increase for low values of P_d. For example, when $P_d = 0{\cdot}5$, the loss is of the order of 6 dB for $a = 0{\cdot}6$.

It is now worthwhile to assess the detection loss suffered by more conventional processing schemes with respect to the new processor of Fig. 4.19. The following three processing schemes have been considered for comparison:

(*a*) binomial MTI with three samples
(*b*) binomial MTI with two samples cascaded with a coherent integrator of three samples
(*c*) binomial MTI with four samples.

The detection losses in Weibull clutter of these three schemes relative to the new processor are listed in Table 4.1.

Table 4.1 *Detection losses of three conventional MTI with respect to the new processor in Weibull clutter (in dB)*

Scheme	*a*		*b*		*c*	
Skewness *a*	1·2	0·6	1·2	0·6	1·2	0·6
$P_d = 0{\cdot}5$	5	14	6·5	12	3	9
$P_d = 0{\cdot}8$	3	9·5	5·5	9	2	5·5

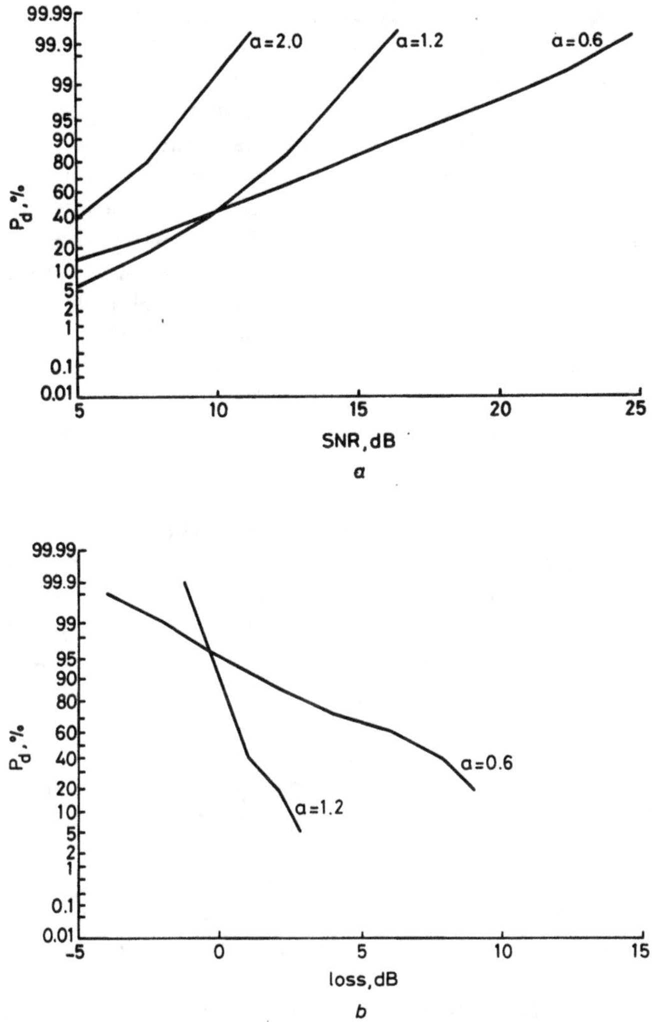

Fig. 4.23 *Detection performance (a) and mismatching loss (b)*
$N=4$, $q=0{\cdot}95$, $P_{fa}=10^{-4}$, CNR = 30 dB, $F_c=0$, $F_s=0{\cdot}5$ PRF, a as parameter

Therefore, it can be noted that considerable losses are suffered for conventional detection schemes when they are fed by Weibull clutter, especially when the skewness parameter is small.

4.3.2 Detector for partially fluctuating target in coherent Weibull clutter

In this case the target signal is not known *a priori*. The target is modelled as a stochastic Gaussian sequence having mean and covariance matrix perfectly known. In more detail, the following models are considered; (i) Swerling 0, corresponding to a signal perfectly known except for the initial phase of the received echo; (ii) Swerling I and II; and (iii) partially fluctuating target echoes encompassing intermediate situations among the two extreme cases of Swerling I and Swerling II. In case (iii) the mean value of the sequence is zero while the ACF is assumed to be Gaussian-shaped. As a consequence, the covariance matrix of the target is characterised by the one-step auto-correlation coefficient q_s. Swerling I is obtained by taking $q_s = 1$, while Swerling II is represented by $q_s = 0$. The detector in Fig. 4.19, where the target $s(k)$ is known *a priori*, is now replaced by a suitable approximation of the optimum target state $\hat{s}_1(k/k-1) = E(s(k)/z(n),\ n = 1, 2, \ldots, k-1;\ H_1)$; the mathematical expression for s is a complicated nonlinear function of Z^{k-1}. An approximation of the signal estimate is obtained by replacing the coherent Weibull clutter (CWC) with an equivalent coherent Gaussian clutter (CGC) with the same mean and covariance matrix values. The H_1 hypothesis becomes

$$z(k) = s(k) + d_G(k) + n(k) \tag{4.8}$$

where $d_G(k)$ is the CGC equivalent to the CWC. The mathematical equations of the mean value $s(k)$ and the corresponding covariance can be found by resorting to least-mean-square estimation theory.

The detection scheme is shown in Fig. 4.24, where the target signal $s(k)$ is estimated from $z(k)$. The covariance $\hat{\Sigma}_s$ of $\hat{s}$ is used to normalise the squared value of the estimation residue $v_1(k/k-1)$.

The detection performance of this detector is shown in Fig. 4.25 for $N = 2$ and in Fig. 4.26 for $N = 3$. Three types of target are considered: the target known *a priori*, target of Swerling 0, and a partially fluctuating target with different values of q_s ($q_s = 1$ is the Swerling I target and $q_s = 0$ is the Swerling II target).

It can be seen from these Figures that the detection performance improves as q_s increases from $q_s = 0$ (Swerling II case) to $q_s = 1$ (Swerling I case). The detection performance for the case of a target known *a priori* and the Swerling 0 case are also shown for comparison.

Fig. 4.27 depicts the detection performance for $P_{fa} = 10^{-6}$ and $a = 0{\cdot}6$. Comparison with Fig. 4.26*c* shows that an increase of approximately 10 dB in SNR is required to obtain approximately the same P_d value but with a lower P_{fa} value.

4.3.3 Adaptive detector for the detection of target embedded in coherent Weibull clutter

In general, the clutter covariance matrix is unknown *a priori*; therefore adaptive detection is desired. An adaptive detector for detecting a target known *a priori* in coherent Weibull clutter was proposed by Farina *et al.* [12]. The principle of the configuration of the adaptive detector is just the same as that of the detectors of Figs. 4.19 and 4.24. The input Weibull clutter is transformed to Gaussian variate, then is added to an adaptive linear prediction

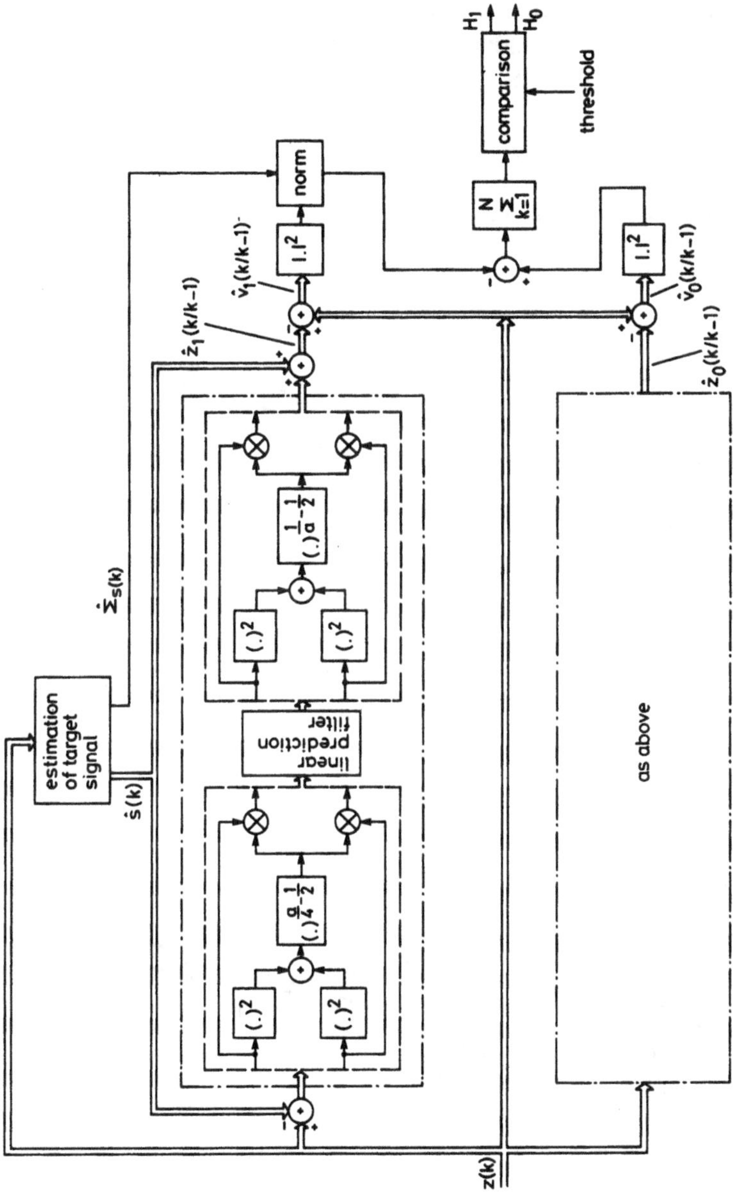

Fig. 4.24 *Detection scheme for a fluctuating target in coherent Weibull clutter*

filter. The adaptivity can be confined to the on-line evaluation of the weights of the linear prediction filter. This is achieved through the estimation of the clutter covariance matrix as seen after the transformation induced by $f^{-1}(\cdot)$. The clutter covariance matrix M_d is evaluated on line by averaging along contiguous range cells around that under test. The (i,j)th element of the matrix is estimated as follows:

$$\hat{M}_d(i,j) = 1/m \sum_{i=1}^{m} z_i(i) z_i^*(j) \tag{4.9}$$

where m is the number of range cells.

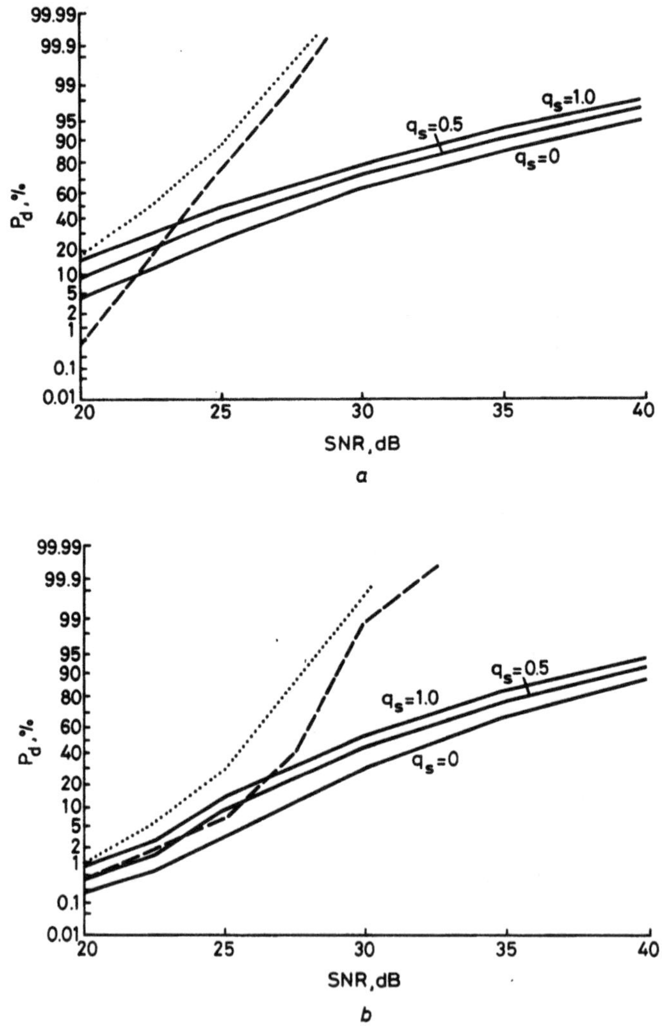

Fig. 4.25

(Continued on opposite page)

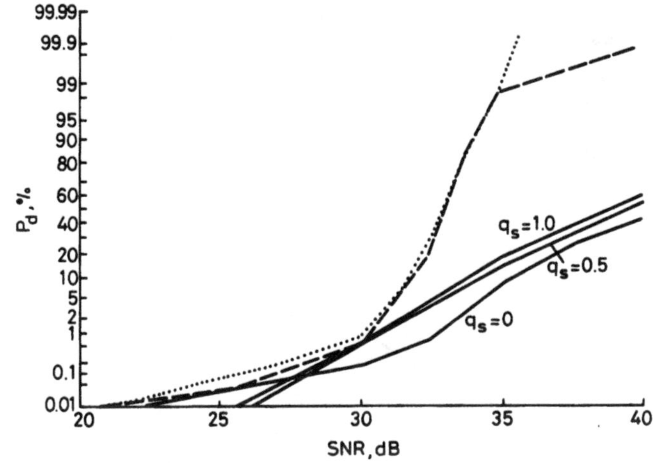

c

Fig. 4.25 *Detection performance of a fluctuating target in CWC*
$N=2$, $q=0{\cdot}95$, $P_{fa}=10^{-4}$, CNR = 30 dB, $F_c=0$, $F_s=0{\cdot}5$ PRF, q_s as parameter
(*a*) $a=2$
(*b*) $a=1{\cdot}2$
(*c*) $a=0{\cdot}6$
.... target known *a priori*
--- Swerling 0
—— partially fluctuating target

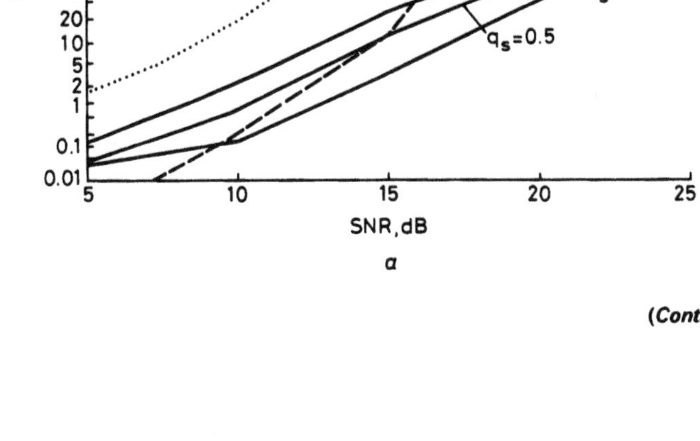

a

Fig. 4.26

(*Continued overleaf*)

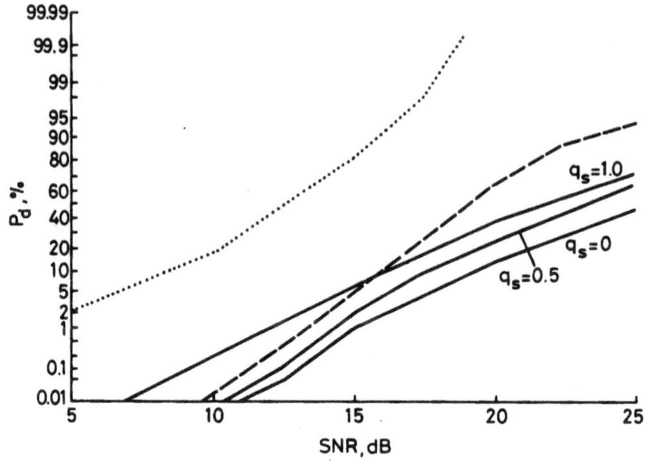

b

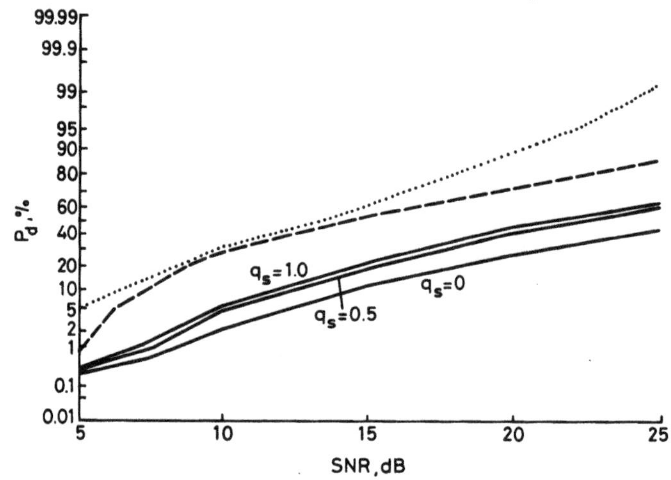

c

Fig. 4.26 *Detection performance of a fluctuating target in CWC*
$N=3$, $q=0{\cdot}95$, $P_{fa}=10^{-4}$, CNR = 30 dB, $F_c=0$, $F_s=0{\cdot}5$ PRF, q_s as parameter
(a) $a=2{\cdot}0$
(b) $a=1{\cdot}2$
(c) $a=0{\cdot}6$
······ target known *a priori*
– – – Swerling 0
—— partially fluctuating

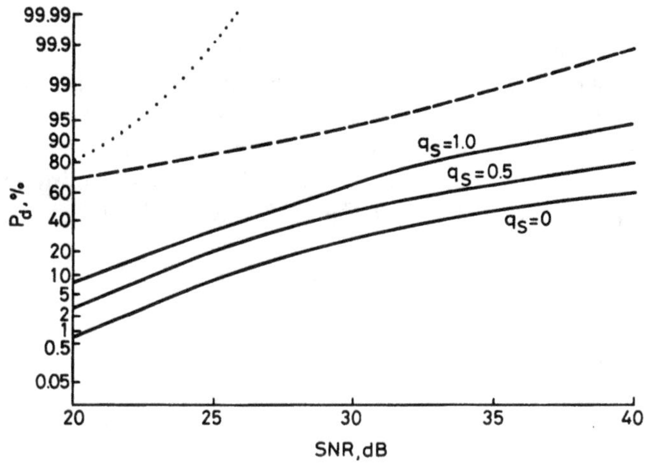

Fig. 4.27 ***Detection performance of a fluctuating target in CWC***
$N=3$, $a=0{\cdot}6$, $q=0{\cdot}95$, $P_{fa}=10^{-6}$, CNR = 30 dB, $F_c=0$, $F_s=0{\cdot}5$ PRF, q as parameter
. . . target known *a priori*
--- Swerling 0
—— partially fluctuating target

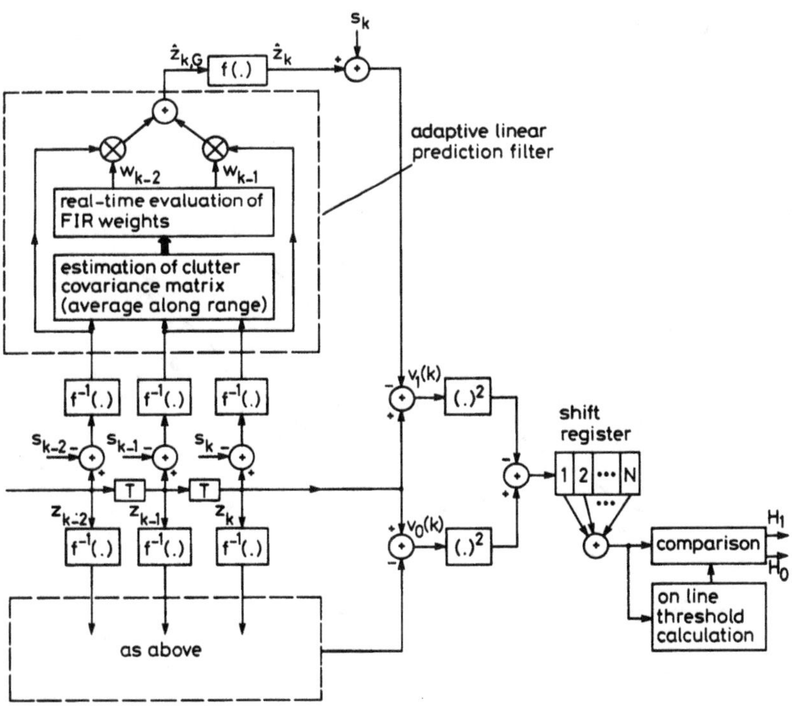

Fig. 4.28 ***Configuration of adaptive detector in Weibull clutter***

The configuration of the adaptive reactor is shown in Fig. 4.28.

The detection loss due to the limited number m of range cells has been evaluated by means of the Monte Carlo simulation technique. Figs. 4.29 and 4.30 show the detection loss for several operational conditions. In particular, Figs. 4.29*a*, *b* and *c* refer to the same number ($m = 10$) of range cells along which the average is performed, the number N of pulses running from 2 to 4. The skewness parameter a is 0·6 for Figs. 4.29*a* and 1·2 for Figs. 4.29*b* and *c*;

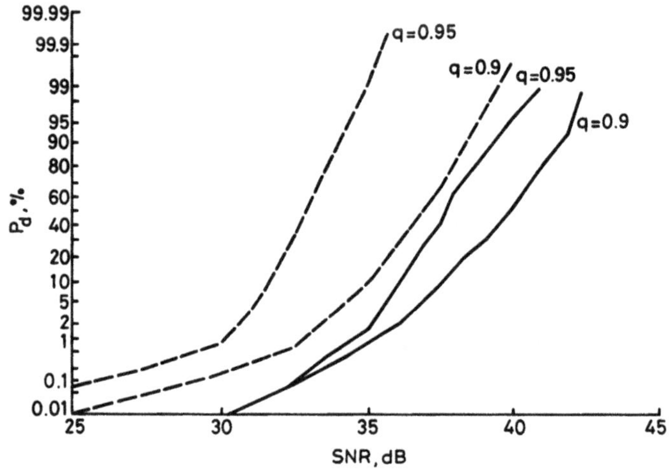

a

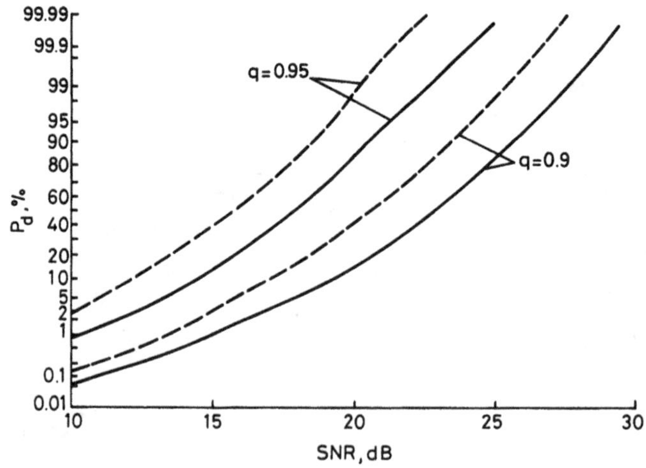

b

Fig. 4.29

(Continued on opposite page)

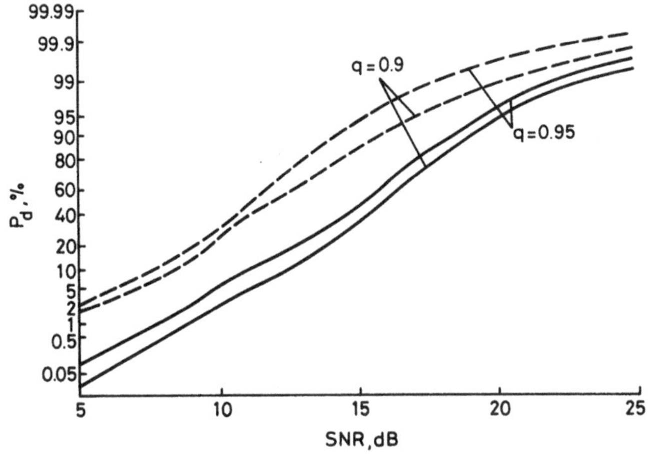

c

Fig. 4.29 *Detection loss due to estimation of filter weights*
$m = 10$, CNR = 30 dB, $F_c = 0$, $F_s = 0{\cdot}5$ PRF, q as parameter
(*a*) $N = 2$, $a = 0{\cdot}6$, $P_{fa} = 10^{-4}$
(*b*) $N = 3$, $a = 1{\cdot}2$, $P_{fa} = 10^{-6}$
(*c*) $N = 4$, $a = 1{\cdot}2$, $P_{fa} = 10^{-6}$
—— adaptive
--- known *a priori*

the probability of false alarm is 10^{-4} for Figs. 4.29*a* and 10^{-6} for Figs. 4.29*b* and *c*.

It can be seen that the detection losses are of the order of 4 dB (for $P_d = 0{\cdot}9$, $q = 0{\cdot}9$, $N = 2$ and $P_{fa} = 10^{-4}$), 2·5 dB (for $P_d = 0{\cdot}9$, $N = 3$, and $P_{fa} = 10^{-6}$) and 5 dB (for $P_d = 0{\cdot}9$, $N = 4$ and $P_{fa} = 10^{-6}$). Fig. 4.30 shows the detection performance of the adaptive detector as a function of the number m of range cells used for averaging purposes.

One of the major limitations of the proposed processors refers to the great number of parameters on which the threshold depends. In addition to P_{fa} and the number of processed pulses N, threshold depends on the clutter correlation coefficient and the clutter/noise values. A method to overcome this problem is to implement a CFAR threshold. The value of the CFAR threshold is found in two steps:

(i) The mean value $\widehat{LLR}$ (log-likelihood ratio) and the standard deviation value σ_{LLR} of the log-likelihood ratio are estimated by averaging along a number of range cells m surrounding the cell under test.
(ii) The detection threshold T is obtained as follows:

$$T = \widehat{LLR} + \gamma \sigma_{LLR} \tag{4.10}$$

where the constant γ depends on the desired P_{fa} value.

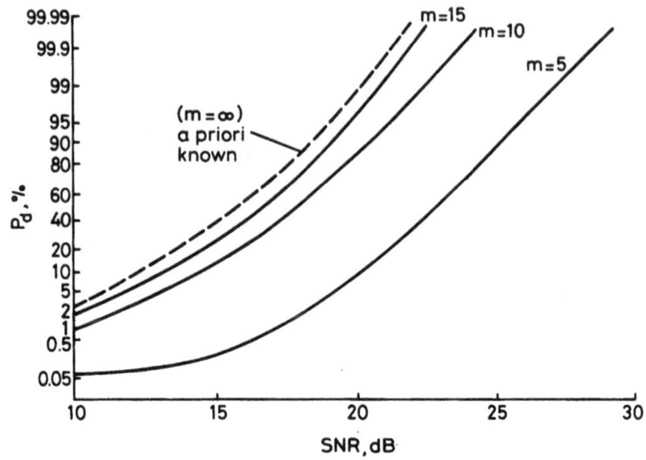

Fig. 4.30 *Detector performance of adaptive detector*

Fig. 4.31 shows the parameter γ against the P_{fa} value. By means of Monte Carlo simulation, it has been shown that the parameter does not change even if the receiver parameters (e.g. CNR, q) are varied. Exception is made for the SNR value (the detector is matched to the target amplitude which is known *a priori*). This is reasonably true if the number of range cells along which the likelihood ratio is averaged is around 10.

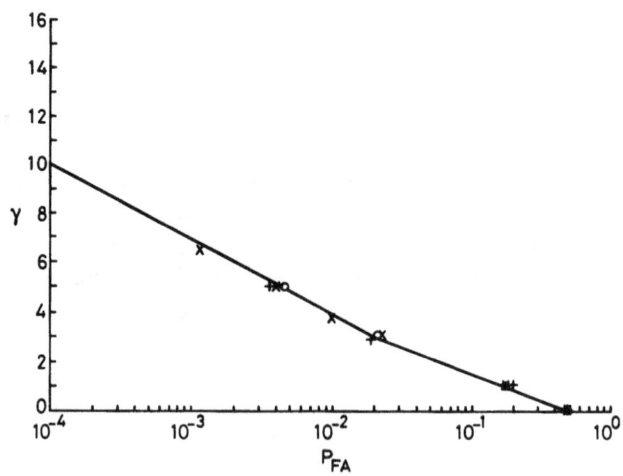

Fig. 4.31 *Parameter γ of CFAR thresholding system*
$N = 2$, $a = 1{\cdot}2$, $m = 10$, CNR = 30 dB,
x $q = 0{\cdot}95$
○ $q = 0{\cdot}9$
+ CNR = 20 dB

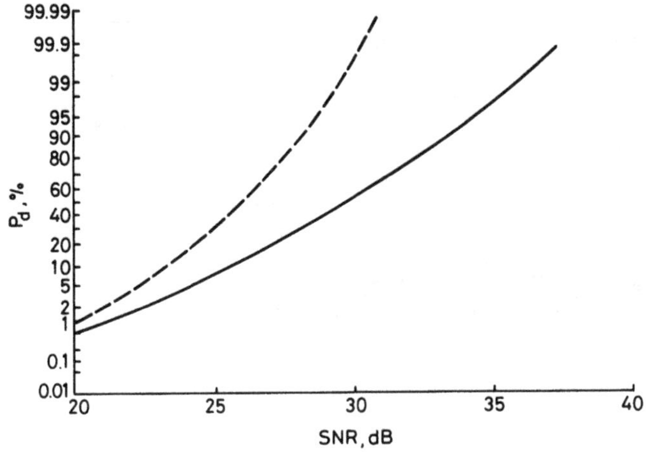

Fig. 4.32 *Detection loss due to CFAR thresholding*
$N = 2$, $q = 0{\cdot}95$, $a = 1{\cdot}2$, $P_{fa} = 10^{-4}$, $m = 10$, CNR = 30 dB, $F_c = 0$, $F_s = 0{\cdot}5$ PFR
--- threshold known *a priori*
—— CFAR threshold

Fig. 4.32 shows the detection loss due to CFAR thresholding for $m = 10$. It is noted that a loss of 5 dB is experienced with 10 range cells when $P_d = 0{\cdot}9$.

The main problem with this adaptive processor is that the transformation from Weibull distribution to Gaussian distribution requires the skewness parameter a to be known *a priori*, which is impossible in practical situation. In other words, the transformation is a parametric one, not a nonparametric process. More loss will occur if the real clutter is not matched to the designed clutter.

4.4 References

1 OLIN, I. D.: 'Characterization of spiky sea clutter for target detection'. IEEE 1984 National Radar Conference, pp. 27–31.
2 BRITTAIN, J. K., SCHROEDER, E. J., and ZEBROWSKI, A. E.: 'Effectiveness of range extended background normalization in ground and weather clutter'. IEE Int. Conf. Radar'77, 1977, pp. 140–144.
3 BLYTHE, J. H. and TRECIOKAS, R.: 'The application of temporal integration to plot extraction', IEE Int. Conf. Radar'77, 1977, pp. 275–279.
4 WHITLOCK, W. S., SHEPHERD, A. M., and QUIGLEY, A. L. C.: 'Some measurements of the effects of frequency agility on aircraft radar returns'. AGARD Conf. Proc. No. 66 on Advanced Radar Systems, 1970, AD–715, p. 485.
5 NATHANSON, F. E., and REILLY, J. P.: 'Radar precipitation echoes', *IEEE Trans.*, 1968, AES–4, pp. 505–514.
6 PIDGEON, V. W.: 'Time, frequency, and spatial correlation of radar sea return', *Space Sys. Planetary Geol. Geophys.*, Americal Astronautical Society, May, 1967; see also Ref. 14.
7 WARD, K. D., and WATTS, S.: 'Radar sea clutter', *Microwave J.*, June 1985, pp. 109–121.

8 WARD, K. D.: 'A radar sea clutter model and its application to performance assessment'. IEE Int. Conf. Radar '82, 1982, pp. 203–207.
9 BAKER, C. J., WARD, K. D., and WATTS, S.: 'The significance and scope of the compound K-distribution model for sea clutter'. IEE Int. Conf. Radar'87, 1987, pp. 207–211.
10 SCHLEHER, D. C.: 'MTI detection performance in Rayleigh and Log-normal clutter'. IEEE 1980 Int. Radar Conf., 1980, pp. 299–304.
11 SCHLEHER, D. C.: 'MTI detection loss in clutter', *Electron. Letts.*, 1981, **17**, 82–83.
12 FARINA, A., RUSSO, A., SCANNAPIECO, F., and BARBAROSSA, S.: 'Theory of radar detection in coherent Weibull clutter', *IEE Proc.* 1987, **134F**, pp. 174–190.
13 FARINA, A., RUSSO, A., and SCANNAPIECO, F.: 'Radar detection in coherent Weibull clutter', *IEEE Trans.*, 1987, **ASSP-35**, pp. 893–895.
14 NATHANSON, F. E.: 'Radar design principles' (McGraw-Hill, 1969), pp. 252–253.

Chapter 5

Appendixes

5.1 Weibull and log-normal distributed sea-ice clutter

Sea-ice clutter was measured using a millimeter-wave radar with a frequency of 35 GHz, antenna beamwidth of 0·25°, vertical beamwidth of 5°, antenna scan rate of 18 rev/min, pulsewidth of 30 ns, pulse-repetition frequency of 4000 Hz, and a transmitted peak power of 30 kW. Data was recorded digitally on a floppy disk as an 8-bit video signal after passing through a log-IF amplifier. One range bin was sampled by 66 data for one pulse and 256 range sweeps were sampled continuously, corresponding to the pulse-repetition frequency. To apply these data to temporal and small-scale range fluctuations, we selected a sample region of 66 range bins and 10 range sweeps corresponding to a beamwidth of about 0·25°. We investigated the Weibull and log-normal distributions using the Akaike Information Criterion (AIC) in appendix 5.2. We obtained the following results.

Range sweep numbers	*Distribution*
0–9	Weibull
10–19	Weibull
20–29	Weibull
30–39	Weibull
40–49	Weibull
50–59	Weibull
60–69	Log-normal
70–79	Weibull
80–89	Weibull
90–99	Weibull
100–109	Weibull
110–119	Weibull
120–129	Weibull
130–139	Weibull

Range sweep numbers	*Distribution*
140–149	Log-normal
150–159	Log-normal
160–169	Weibull
170–179	Weibull
180–189	Log-normal
190–199	Weibull
200–209	Weibull
210–219	Log-normal
220–229	Weibull
230–239	Weibull
240–249	Weibull

Thus most of sea-ice clutter obeys a Weibull distribution.

Range-bin numbers against range-sweep numbers for sea-ice clutter are shown in Fig. 5.1. Circle means a target of iron tower with the height of 15 m above the sea surface.

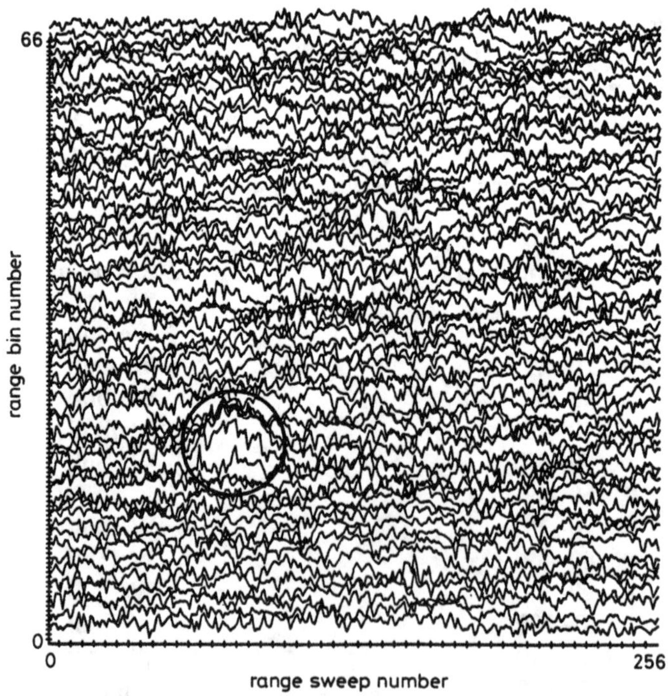

Fig. 5.1 *Range bin number against range sweep number for sea-ice clutter*

5.2 Akaike Information Criterion

The Weibull probability density function is written as follows:

$$p_c(x) = \begin{cases} \dfrac{c}{b}\left(\dfrac{x}{b}\right)^{c-1} e^{-(x/b)^c} & \text{for } x > 0,\ b > 0 \text{ and } c > 0 \\ 0 & \text{otherwise} \end{cases} \tag{5.1}$$

Here x is the amplitude of the return signals, b is a scale parameter and c is a shape parameter.

Eqn. 5.1 is integrated to obtain

$$Y = cX - c \ln b \tag{5.2}$$

where

$$Y = \ln\left[-\ln\left\{1 - \int_0^x p_c(x)\, dx\right\}\right] \tag{5.3}$$

and

$$X = \ln x \tag{5.4}$$

From eqn. 5.2, the shape parameter c is easily estimated from a plot of Y against X.

The log-normal probability density function is written as follows:

$$p(x) = \frac{1}{\sqrt{2\pi}\sigma x} e^{-(1/2\sigma^2)(\ln x - m)^2}$$

$$= \frac{1}{\sqrt{2\pi}\sigma x} e^{-(1/2\sigma^2)(\ln x/x_m)^2} \tag{5.5}$$

where x is the amplitude of the radar return signals. x_m is the median value of x and σ is the standard deviation of $\ln(x/x_m)$.

Eqn. 5.5 is integrated to obtain

$$Y = \frac{1}{\sigma} X - \frac{1}{\sigma} \ln x_m \tag{5.6}$$

where

$$Y = \operatorname{erf}^{-1}\left[2 \int_0^x p(x) - 1\right] \tag{5.7}$$

and

$$X = \ln x \tag{5.8}$$

From eqn. 5.6, the log-normal-distribution model is easily estimated from a plot of Y against X. Using the sea-ice clutter data in Appendix 5.1, one example for range sweep numbers 0—9 is shown in Figs. 5.2 and 5.3. Thus the number of data points is 660.

In Figs. 5.2 and 5.3, a straight line was fitted to the values of Y and X by the least-squares method. If the data follow a Weibull distribution or a log-normal distribution, they lie on a straight line in this representation, and the slope gives the shape parameter c in the Weibull distribution and the parameter σ in the log-normal distribution. The root-mean-square error (RMSE) is the deviation of the data points from the straight line drawn by the least-squares methods. The smaller values of RMSE mean a good fit to the distribution. As seen from Figs. 5.2 and 5.3, a Weibull distribution is a better fit than a log-normal distribution. As an alternative to this approach, we consider the Akaike Information Criterion which is a rigorous fit of the distribution to the data.

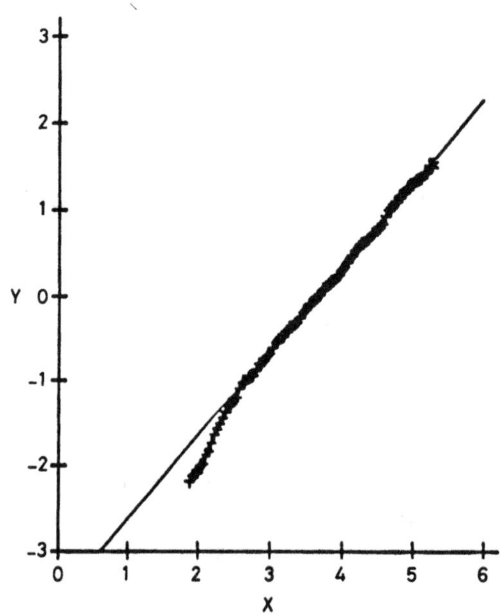

Fig. 5.2 *Determination of c for a Weibull distribution from range sweep numbers 0–9* $c = 0{\cdot}98$, $b = 39{\cdot}4$ *and* RMSE = 0·024

Data file	00—07.t2
Line	0—9
Distance	25—90
Amp of Dot	49—137
Amp of LSM	70—137
b parameter	39·373517
c parameter	0·980939
RMSE	0·024093

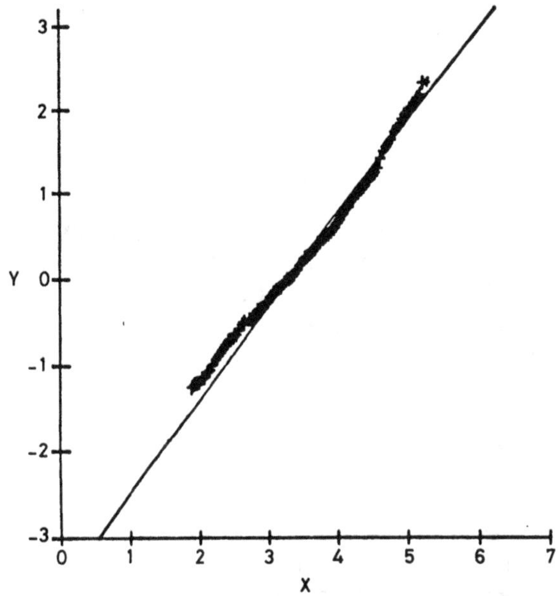

Fig. 5.3 *Determination of σ for a log-normal distribution from range sweep numbers 0–9*
$\sigma = 0{\cdot}92$, $m = 3{\cdot}29$ *and* RMSE = 0·094

Data file	00—07.t2
Line	0—9
Distance	25—90
Amp of Dot	49—137
Amp of LSM	70—137
Mean	3·289003
Sigma	0·919332
RMSE	0·093726

Considering two or more hypothetical probability distribution models, we have a criterion by which the model gives an optimum fit to the data. This is called Akaike Information Criterion which is abbreviated by AIC.* The AIC is explained as follows:

First we assume that the true probability distribution $(p_1, p_2, \ldots, p_N)$ is known. Here p_N is the probability that the Nth event occurs. Next we will consider a sufficiently large number of trials. Then the Nth event will occur approximately $m = Mp_N$ times. As a model, we assume the probability distribution $(q_1, q_2, \ldots, q_N)$. By observing the M samples obeying this distribution, the probability W is written as

$$W = \frac{M!}{m_1!, m_2!, \ldots, m_N!} q_1^{m_1} q_2^{m_2} \cdots q_N^{m_N} \tag{5.9}$$

*AKAIKE, H.: 'Information theory and an extension of the maximum likelihood principle', *in* PETROV, B. N., and CSAKI, F. (Eds.), 2nd International Symposium on Information Theory (Akademiai Kiado, Budapest, 1973) pp. 26–281

Here W is the probability that we obtain the probability distribution. $(m_1, m_2, \ldots, m_N)$. By taking a logarithm of both sides of eqn. 5.9 and dividing by M, we obtain

$$\frac{\ln(W)}{M} \underset{M\to\infty}{\longrightarrow} B(p, q) = p_N \ln\left(\frac{q_N}{p_N}\right) \tag{5.10}$$

Here $B(p, q)$ is called Kullback–Leibler's entropy.† From the above discussion, the probability is that the predicted distribution realized becomes large with larger values of B. In this sense, B is used as a model estimation; i.e. the larger values of B mean a good model.

The Kullback–Leibler entropy is rewritten as

$$B = \sum p_N \ln(q_N) - \sum p_N \ln(p_N) \tag{5.11}$$

The second term on the right-hand side depends only on a true distribution. Therefore, only the first term plays an important role in estimating the model. This term is interpreted as an expected value of $\ln(q_N)$. Therefore, the first term is estimated from the M numbers of the observed values $x_1, x_2, \ldots, x_M$. Then the logarithmic likelihood L is defined as

$$L = \sum \ln\{f(x_N)\}, \quad f(x_k) = q_N \quad \text{for } x_k = N \tag{5.12}$$

Here a function $f(x)$ is a probability that the observed values are x, and depends on the model. The larger L is the better model. Now we assume that the probability density function $f(x)$ has parameters θ. Then we can write the probability density function model as $f(x\colon \theta)$ for a stochastic variable x and parameters θ. If θ have k numbers of parameter, then θ are k-dimensional vectors. In this case, the logarithmic likelihood $L(\theta)$ is defined as

$$L(\theta) = \sum \ln\{f(x\colon \theta)\} \tag{5.13}$$

Eqn. 5.12 is determined from the observed values. However, if $f^*(x)$ is a true probability density function, then the true logarithmic likelihood is written as

$$L^*(\theta) = M \int f^*(x) \ln\{f(x\colon\theta)\}\, dx \tag{5.14}$$

Usually, $L^*(\theta)$ cannot be calculated, as long as the true probability density

†KULLBACK, S., and LEIBLER, R. A.: 'On information and sufficiency', *Ann. Math. Statist.*, 1953, **22**, pp. 79–86

function is not known. However, it is well known that $L(\theta_0) - k$ is an unbiased estimation of the logarithmic likelihood $L(\theta_0)$. Here θ_0 is the maximum likelihood estimation to obtain the largest $L^*(\theta)$. Therefore, finally, the AIC for a given model is defined as

$$\begin{aligned}\text{AIC} &= -2[\{\text{maximum logarithmic likelihood}\} \\ &\quad -\{\text{number of parameters included in the model}\}] \\ &= -2\{L(\theta_0) - k\} \qquad (5.15)\end{aligned}$$

The model which yields the smallest AIC is regarded as the best one.

5.3 Determination of the optimum probability-density function for sea-ice clutter using AIC

The results are shown in Figs. 5.4*a*—*y* from range sweep numbers 0 to 245. The final results are summarised in Appendix 5.1. The optimum probability distribution is determined by choosing the smallest AIC value.

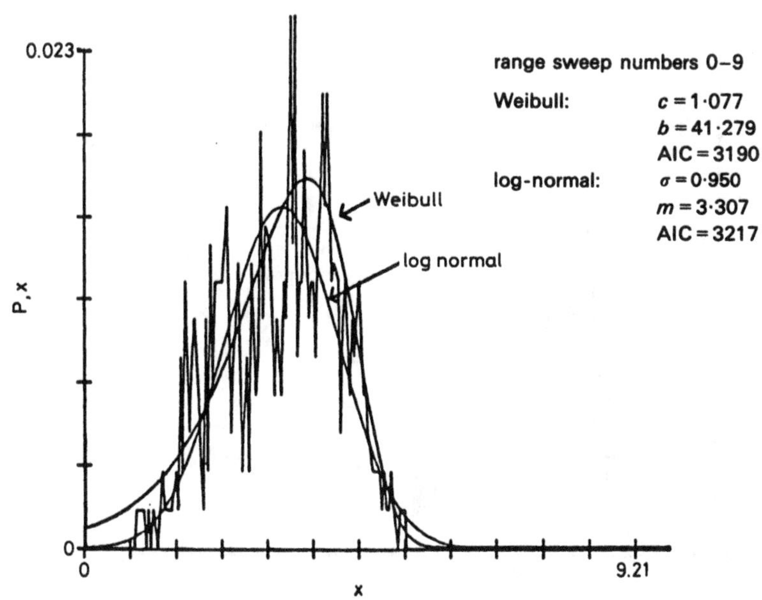

a

Fig. 5.4a *Weibull distribution is a better fit to the data*

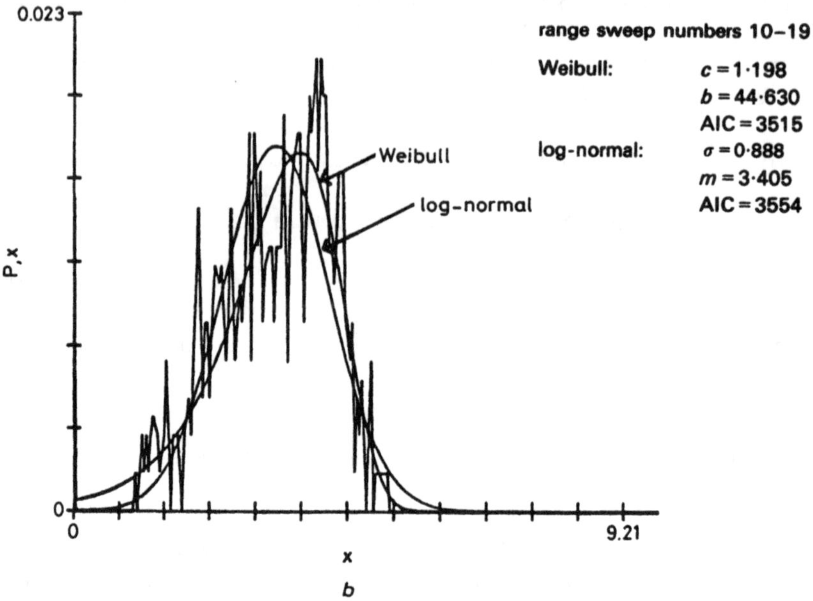

Fig. 5.4b *Weibull distribution is a better fit to the data*

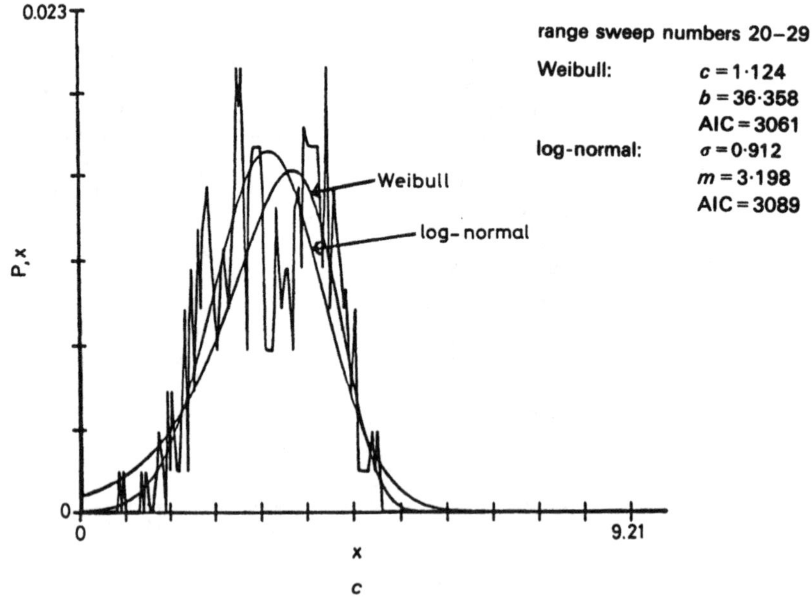

Fig. 5.4c *Weibull distribution is a better fit to the data*

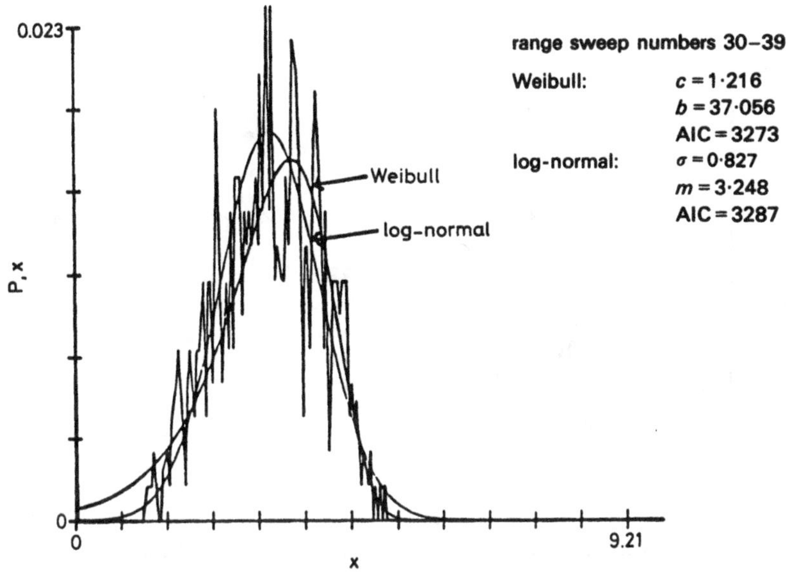

d

Fig. 5.4*d* *Weibull distribution is a better fit to the data*

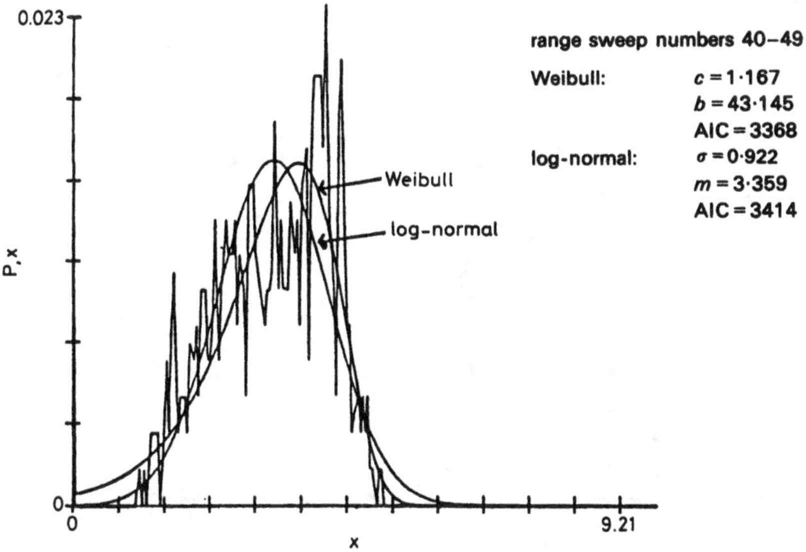

e

Fig. 5.4*e* *Weibull distribution is a better fit to the data*

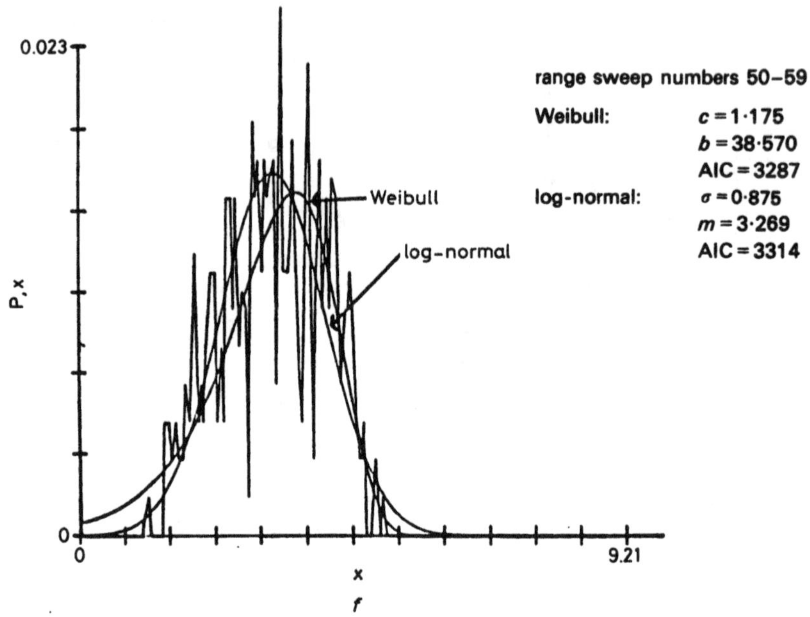

Fig. 5.4f *Weibull distribution is a better fit to the data*

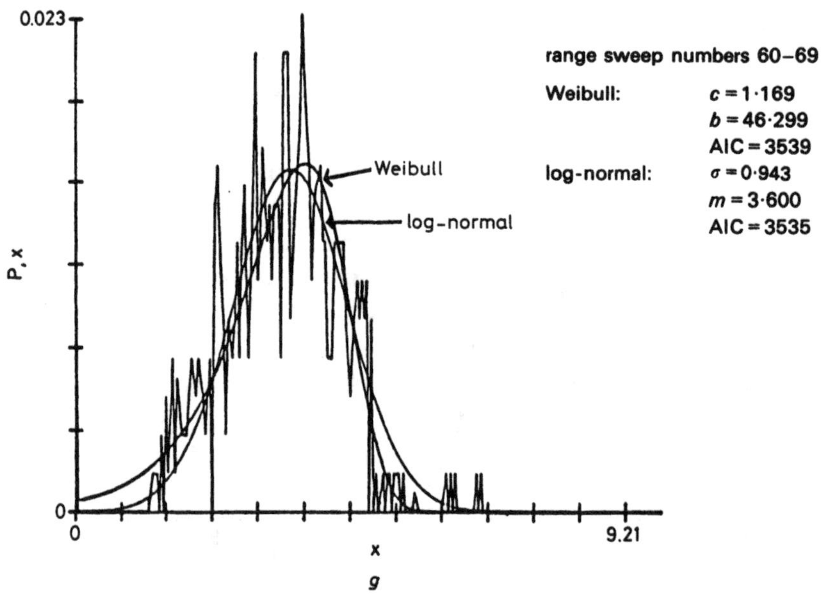

Fig. 5.4g *Log-normal distribution is a better fit to the data*

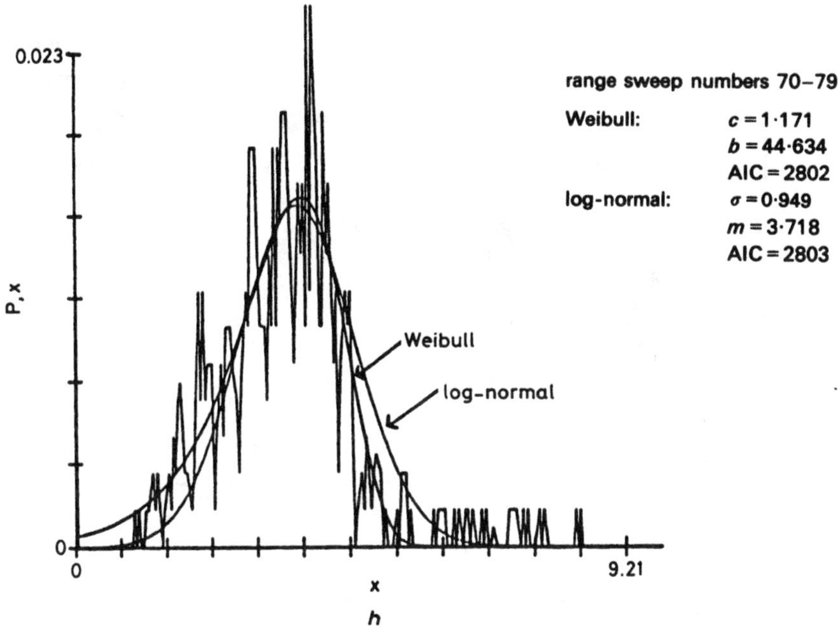

Fig. 5.4*h* *Weibull distribution is a better fit to the data*

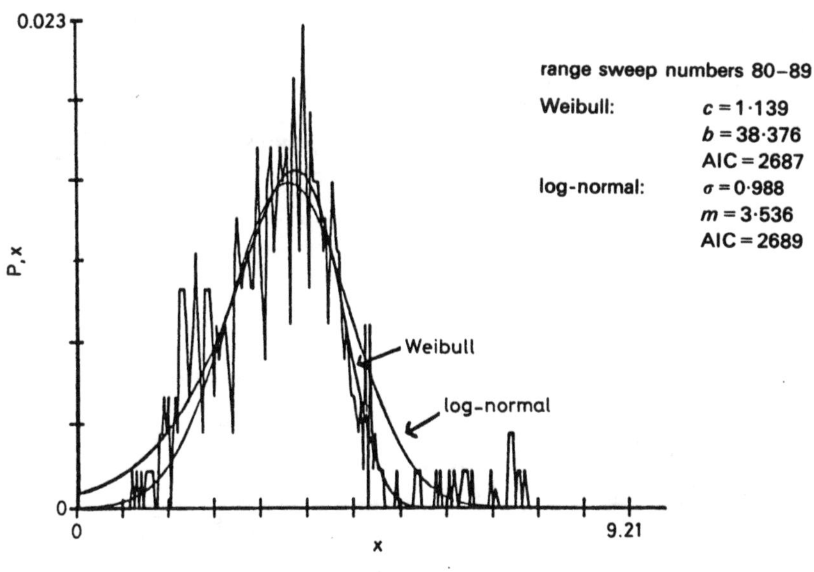

Fig. 5.4*i* *Weibull distribution is a better fit to the data*

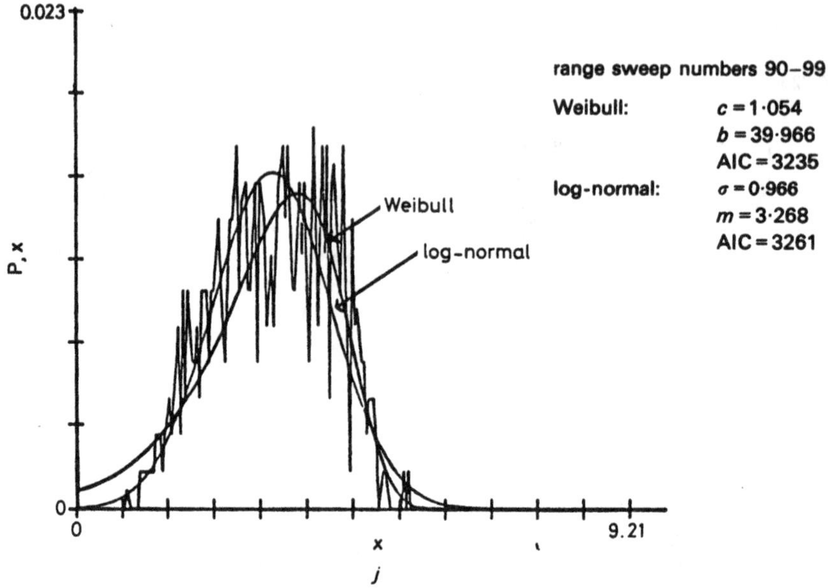

Fig. 5.4*j* *Weibull distribution is a better fit to the data*

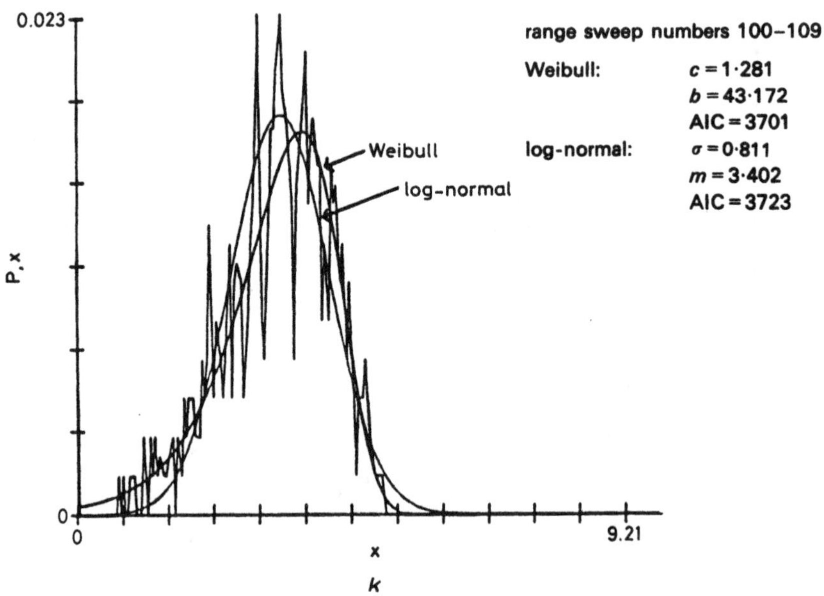

Fig. 5.4*k* *Weibull distribution is a better fit to the data*

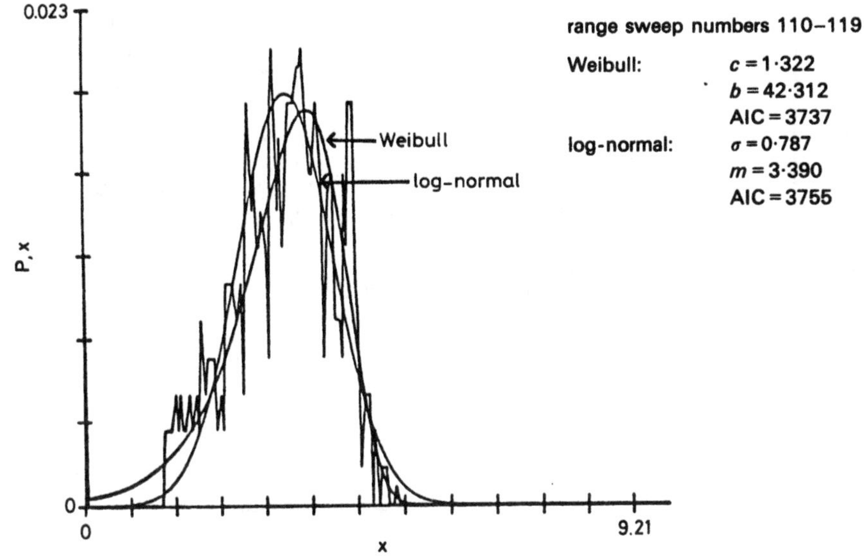

l

Fig. 5.4*l* *Weibull distribution is a better fit to the data*

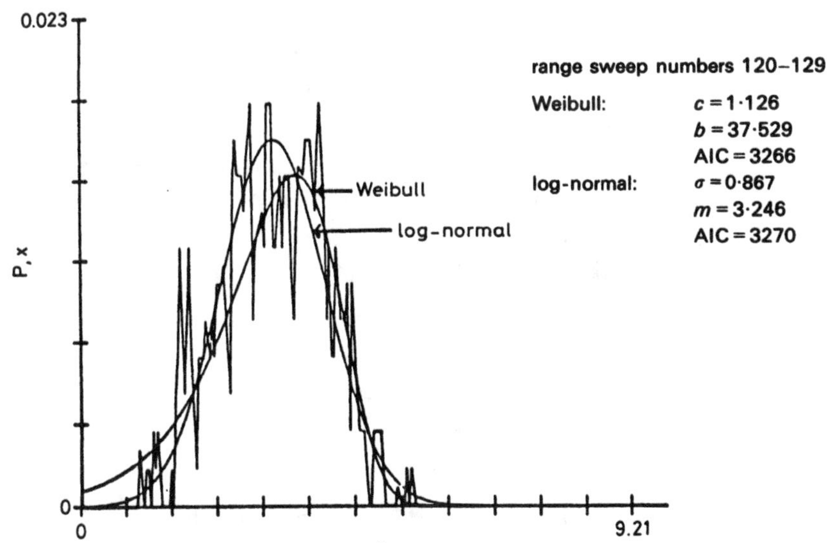

m

Fig. 5.4*m* *Weibull distribution is a better fit to the data*

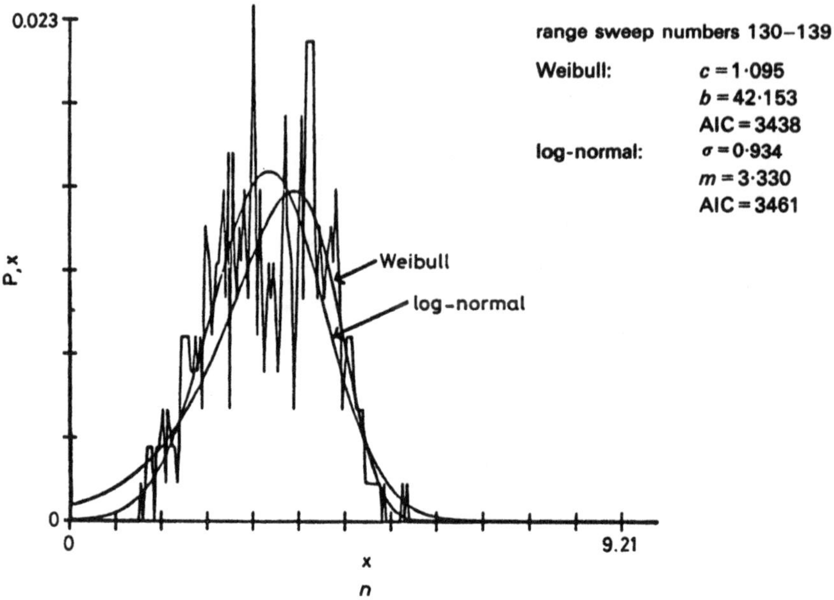

Fig. 5.4*n* *Weibull distribution is a better fit to the data*

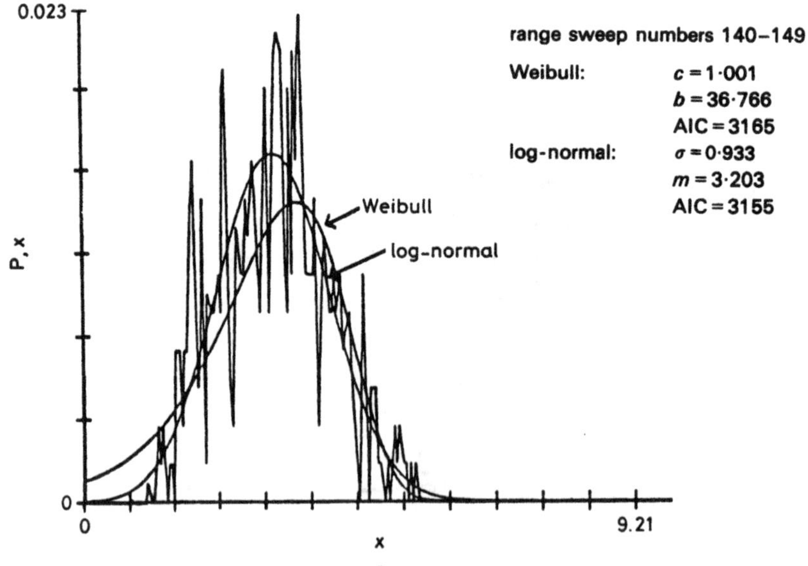

Fig. 5.4*o* *Log-normal distribution is a better fit to the data*

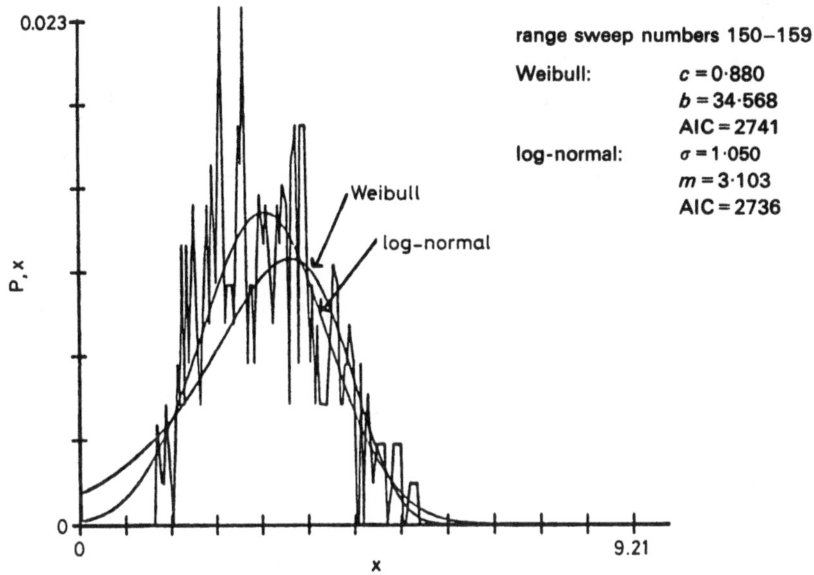

p

Fig. 5.4*p* *Log-normal distribution is a better fit to the data*

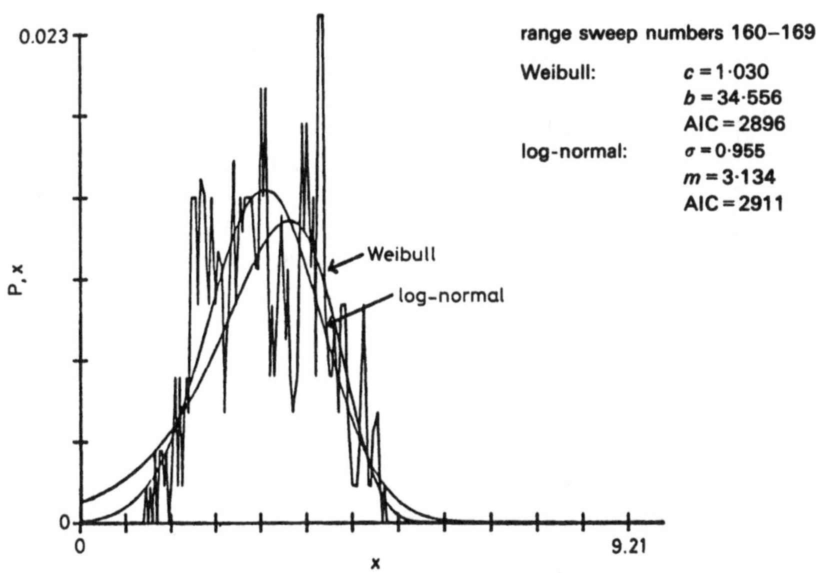

q

Fig. 5.4*q* *Weibull distribution is a better fit to the data*

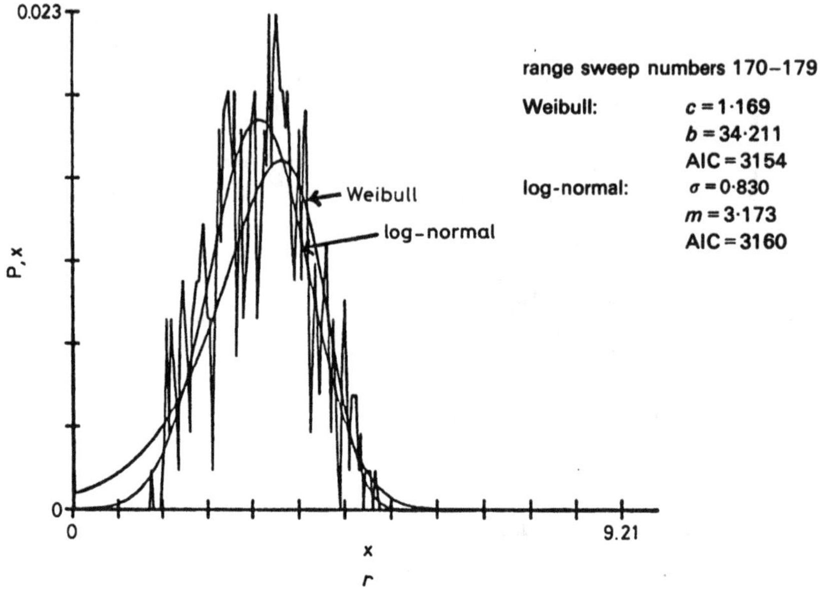

Fig. 5.4r *Weibull distribution is a better fit to the data*

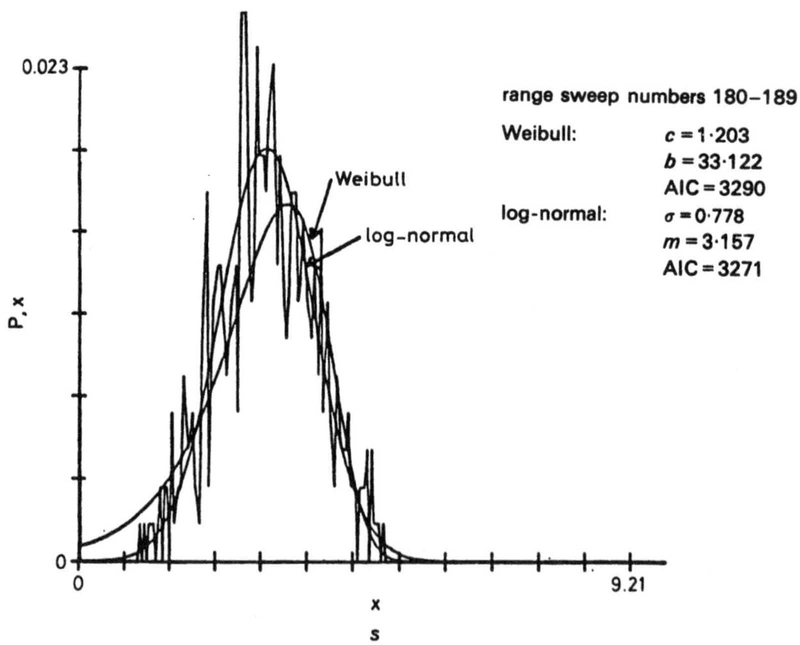

Fig. 5.4s *Log-normal distribution is a better fit to the data*

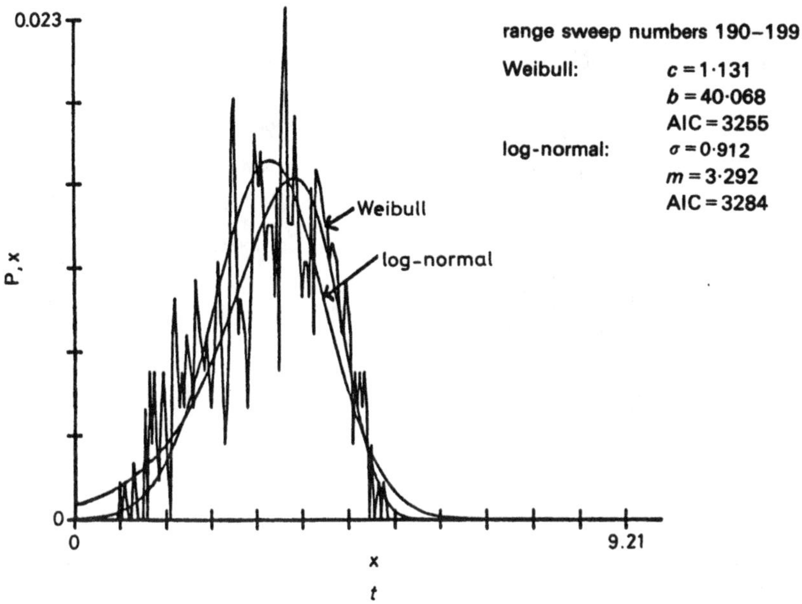

Fig. 5.4t *Weibull distribution is a better fit to the data*

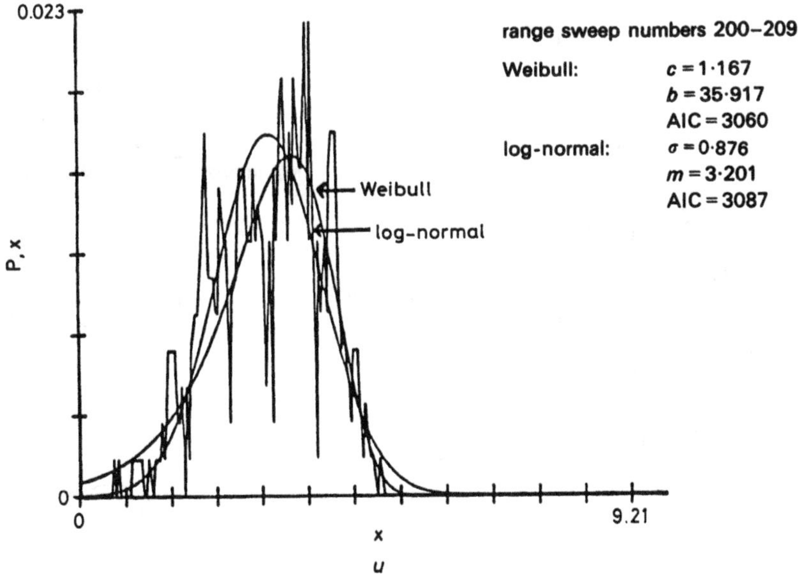

Fig. 5.4u *Weibull distribution is a better fit to the data*

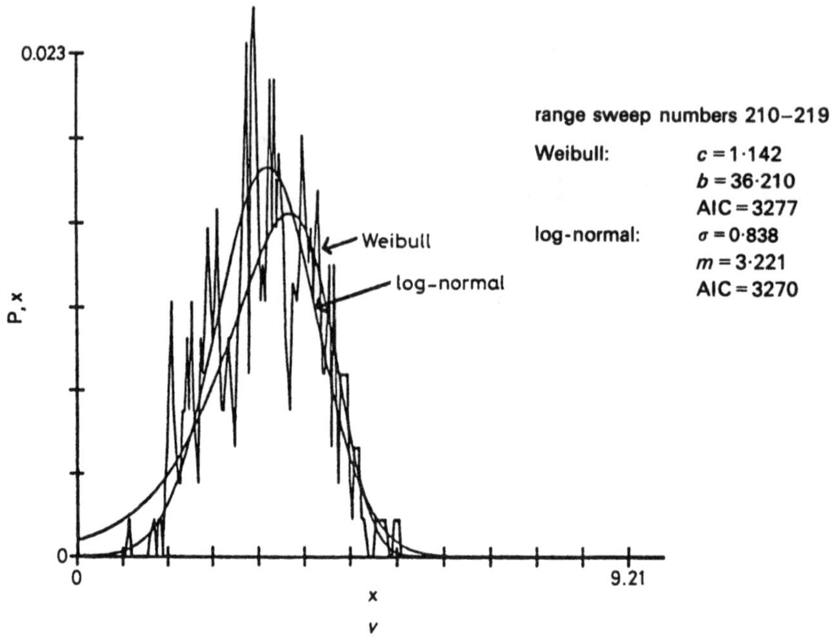

Fig. 5.4v *Log-normal distribution is a better fit to the data*

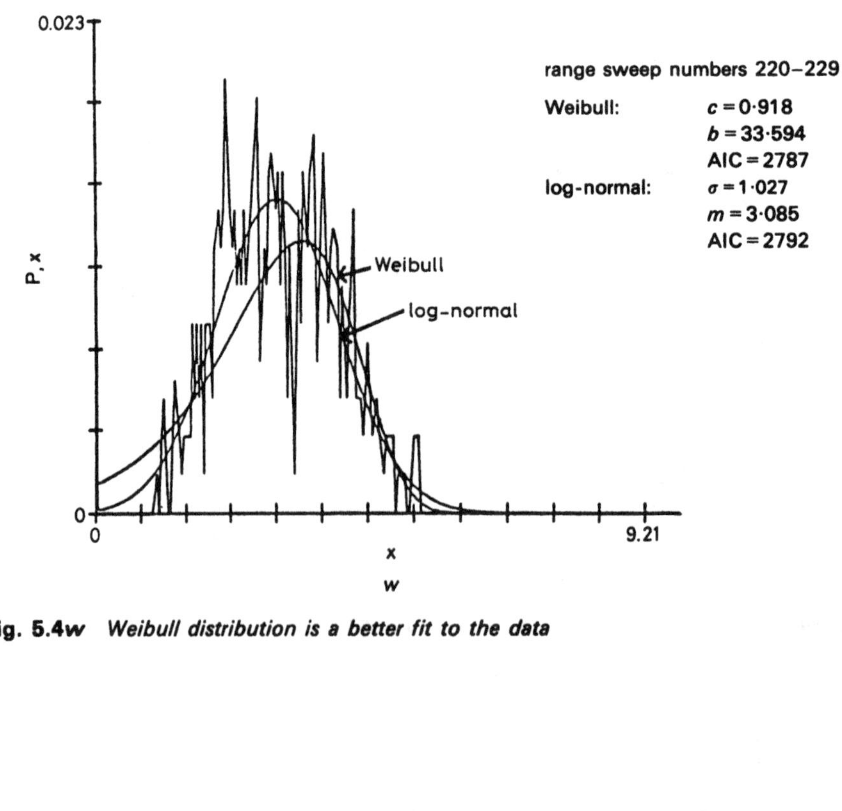

Fig. 5.4w *Weibull distribution is a better fit to the data*

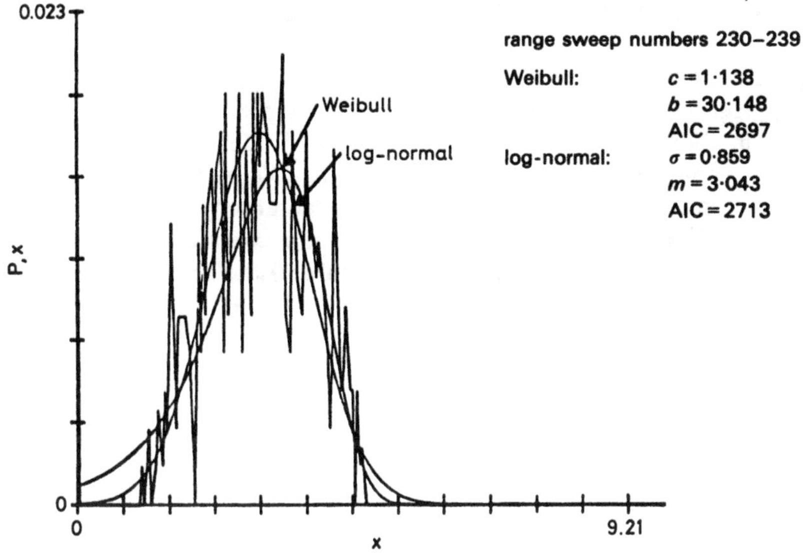

x

Fig. 5.4x ***Weibull distribution is a better fit to the data***

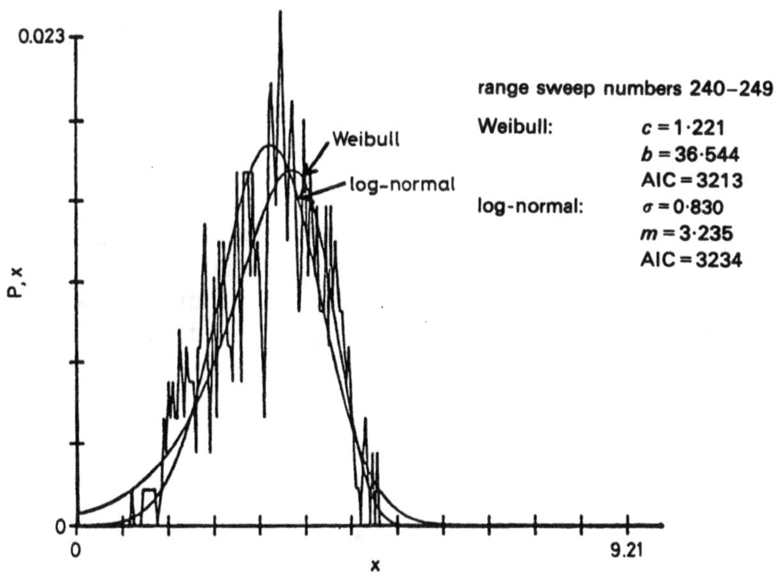

y

Fig. 5.4y ***Weibull distribution is a better fit to the data***

Fig. 5.4 ***Determination of optimum probability density function using AIC from sea-ice clutter. The smallest value of AIC is the optimum probability density function.***

5.4 Suppression of Weibull sea-ice clutter and detection of target

We have found that sea-ice clutter obeys almost a Weibull distribution. Here we apply Hansen's method in the text to the suppression of sea-ice clutter. Hansen's method is based on a Weibull CFAR detector that takes into account the nonlinear transformation from the Weibull to the exponential probability-density function. This method is generalised as follows:

Let the amplitude of Weibull clutter be x and y be its output after passing through a logarithmic amplifier. Then the first and the second moments of y are given by

$$\langle y \rangle = \int_0^\infty \ln(x) p_c(x)\, dx = \ln(b) - \frac{\gamma}{c} \tag{5.16}$$

$$\langle y^2 \rangle = \int_0^\infty \ln^2(x) p_c(x)\, \mathrm{d}x = \left(\ln(b) - \frac{\gamma}{c} \right)^2 + \frac{\pi^2}{6c^2} \tag{5.17}$$

where $\gamma = 0{\cdot}5772\ldots$ is Euler's constant. The Weibull probability-density function $p_c(x)$ is written in eqn. 5.1. The variance of y is derived from eqns. 5.16 and 5.17 as

$$\mathrm{var}(y) = \langle y^2 \rangle - \langle y \rangle^2 = \frac{\pi^2}{6c^2} \tag{5.18}$$

The variance of y depends only on the c value of the shape parameter. The c value is found from eqn. 5.18 and the b value is found from eqn. 5.16. Thus it is necessary to determine two Weibull parameters, c and b values, by using a finite number of data samples passed through a logarithmic amplifier. Now a new variable z is introduced as

$$\left(\frac{x}{b} \right)^c = z^m \tag{5.19}$$

where m is an arbitrary constant. From eqns. 5.1 and 5.19, it is easily seen that the variable z obeys the following distribution:

$$p(z) = m z^{m-1} e^{-z^m} \tag{5.20}$$

This distribution is independent of the shape and scale parameters of the input

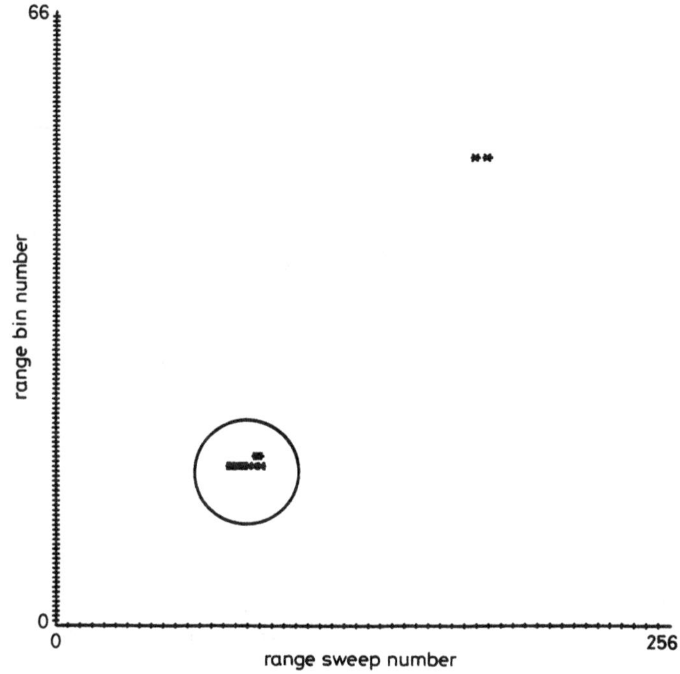

Fig. 5.5 *Suppression of sea-ice clutter for a finite number of data samples 16 and false-alarm probability* 10^{-5}

signals. Thus CFAR is obtained. For $m = 1$, eqn. 5.20 is identical to an exponential distribution proposed by Hansen. Now we will transform to a Rayleigh distribution of $m = 2$ using observed sea-ice clutter data. A finite number of data samples 16 and false-alarm probability 10^{-5} were considered. The result is shown in Fig. 5.5. By comparing with original Fig. 5.1, it is easily seen that sea-ice clutter was suppressed and the target was detected.

Index

www.ingramcontent.com/pod-product-compliance
Lightning Source LLC
LaVergne TN
LVHW011802200926
848484LV00003B/765

* 9 7 8 0 8 6 3 4 1 1 9 1 5 *